Cogeneration of Electricity and Useful Heat

Editors

Bruce W. Wilkinson, Ph.D.
Professor
Department of Chemical Engineering
Michigan State University
East Lansing, Michigan

Richard W. Barnes, M.Sc.
Senior Researcher
Department of Energy
Oak Ridge National Laboratory
Oak Ridge, Tennessee

CRC Press, Inc.
Boca Raton, Florida

Library of Congress Cataloging in Publication Data

Main entry under title:

Cogeneration of electricity and useful heat.

  Bibliography: p.
  Includes index.
  1.   Electric power production. 2. Heat
engineering. I. Wilkinson, Bruce Wendell,
1928-   II.   Barnes, Richard W.
TK1005.C6   621.4'028   79-11770
ISBN O-8493-5615-6

   Direct all inquiries to CRC Press, Inc., 2000 N.W. 24th Street, Boca Raton, Florida, 33431.

© 1980 by CRC Press, Inc.

International Standard Book Number 0-8493-5615-6

Library of Congress Card Number 79-11770
Printed in the United States

THE EDITORS

Bruce W. Wilkinson, Ph.D., is Professor of Chemical Engineering at Michigan State University, East Lansing, Michigan. He received his BChE, MSc and Ph.D. degrees in Chemical Engineering from the Ohio State University in 1951 and 1958.

Dr. Wilkinson has been on the faculty at Michigan State University since 1965. Prior to that time, he was employed in research, process design and project engineering for the Dow Chemical Co. for 11 years. He is a member of the American Institute of Chemical Engineers, American Nuclear Society and American Society for Engineering Education. He has been actively involved in the nuclear field and in the examination of energy alternatives.

Richard W. Barnes is a Senior Research Staff Member at the Oak Ridge National Laboratory where he is responsible for studies dealing with the modeling and analysis of industrial energy use in the United States.

Prior to coming to Oak Ridge National Laboratory in 1978, Mr. Barnes was with the Dow Chemical Company for 15 years and with the Shell Oil Company for 16 years. He has held positions in research, production, corporate administration and management. Mr. Barnes holds a BSc degree in Chemical Engineering from the University of Utah, an MSc in Chemical Engineering from the University of Washington, and an MSc in Industrial Administration from the University of Michigan. He is a member of the Operations Research Society of America and has concentrated much of his working experience in industrial systems design and analysis.

CONTRIBUTORS

John P. Ackerman, Ph.D.
Technical Manager
Department of Energy
Molten Carbonate Fuel Cell
Systems Development Program
Argonne National Laboratory
Argonne, Illinois

Peter G. Bos
Resource Planning Associates, Inc.
Cambridge, Massachusetts

William S. Butler, P.E.
Energy Consultant
Riddick Engineering Corporation
Little Rock, Arkansas

Jerry P. Davis, M.S.Nuc.E.
President
Energy Systems Division
Thermo Electron Corporation
Waltham, Massachusetts

Norman L. Dean, Jr., J.D.
Attorney
Washington, D.C.

Martin C. Doherty, M.S.M.E.
Consulting Applications Engineer
Thermal Power Systems
General Electric Company
Schenectady, New York

Charles D. Glass
Vice President
Gulf States Utilities Company
Beaumont, Texas

J. Robert Hamm, B.S.M.E.
Advisory Engineer
Westinghouse Research Laboratories
Pittsburgh, Pennsylvania

Paul T. Hodiak, M.S.M.E.
General Manager
Applied Energy, Inc.
San Diego, California

Robert E. Holtz, Ph.D.
Heat Utilization Section Manager
Components Technology Division
Argonne National Laboratory
Argonne, Illinois

Michael A. Karnitz, Ph.D.
Group Leader
Community Systems Programs
Energy Division
Oak Ridge National Laboratory
Oak Ridge, Tennessee

William R. Mixon, M.S.Eng.
Section Head
Energy Conservation and Use Analysis
Oak Ridge National Laboratory
Oak Ridge, Tennessee

Mitchell Olszewski, Ph.D.
Research Staff Member
Engineering Technology Division
Oak Ridge National Laboratory
Oak Ridge, Tennessee

Thomas E. Root, B.S.
Senior Engineer
Planning Department
Detroit Edison Company
Detroit, Michigan

James H. Williams, Ph.D.
Senior Associate
Resource Planning Associates, Inc.
Washington, D.C.

ACKNOWLEDGMENT

The cooperation and support of the Dow Chemical Company and Michigan State University is gratefully recognized.

TABLE OF CONTENTS

Chapter 1

INTRODUCTION AND OVERVIEW

R. W. Barnes

TABLE OF CONTENTS

I. INTRODUCTION

This book is intended to serve the needs of practicing engineers, economic analysts, policymakers, and managers in comparing the status and characteristics of cogeneration alternatives. It identifies the current and potential future approaches to cogeneration and reviews the significant technical and economic aspects of each. It will enable the reader to determine the applicability of the concept to a particular situation and to initiate the planning of a chosen system. A bibliography is included for each section to enable the reader to delve into a more detailed analysis, if desired. Presentations include discussion of the effects of fuel alternatives, load dynamics, technology trends, and other parametric factors which bear upon the optimum energy system.

In the following pages of this introduction, basic concepts of cogeneration are briefly discussed to provide a foundation for the nontechnical reader.

There are three principal approaches to cogeneration: central utility plant systems, industrial plant systems, and total energy or modular integrated utility systems (MIUS). While the major attributes of each of these forms of cogeneration are distinguishable, definitions of each are imprecise, and much overlap exists in applications, as suggested in Figure 1. While this three-way categorization is used in subsequent chapters, the reader will in several places find similar material treated under more than one heading.

The term *cogeneration* characterizes energy conversion processes in which heat is generated for a dual purpose, usually to produce *both* electricity and a flow of otherwise useful heat. The heat output is conventionally in the form of steam or hot water. Such dual-purpose processes can substantially improve the efficiency of energy utilization for electrical power generation. In the following sections of this chapter, various basic concepts of cogeneration are discussed. The reader with some knowledge of cogeneration may want to skip over this material.

II. ELECTRICAL POWER GENERATION

Essentially all of the electricity used in the U.S. is produced in electrical generators. They range in size from one half kilowatt units (used in automobiles) to the million kilowatt units used by the utility companies to supply centrally distributed electricity for homes, businesses, and industry. By far the most widely used power source for generation operation is the steam turbine, which uses the energy present in a flow of high pressure steam to produce the mechanical power needed to turn the rotor of a generator. Steam to operate the turbine is produced when fuels are consumed. The consumed fuel both evaporates water in a boiler and raises the resulting steam temperature and pressure to the required operating levels. The higher the steam temperature, the more energy it will carry and the more work it will do in the turbine. Over the years, as technology has improved, typical steam properties for new utility power stations have increased from 900°F and 1200 pounds of pressure per square inch (psi) 30 years ago, to 1000°F and 2400 psi today. This improvement alone has reduced the fuel input needed to produce a unit of electricity by 15%.

Combustion turbines (Figure 2), diesel, and gasoline engines are also used for power generation. In these machines, fuel combustion occurs *within* the turbine or engine, and the pressures created by the combustion gases are converted directly to mechanical work transmitted through a drive shaft to the generator. Because combustion turbines, diesel, and gasoline engines use internal combustion, they often function at higher temperatures than do steam turbines and should theoretically be even more efficient in producing work from a given fuel-energy input. In practice, however, the internal

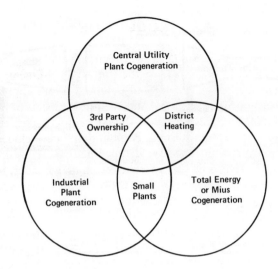

FIGURE 1.   Forms of cogeneration.

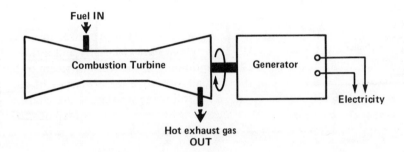

FIGURE 2.   Combustion turbine for power generation.

combustion devices usually exhaust their combustion gases at such high temperatures that much of the input energy passes through unused, and the potential efficiency advantage is lost.

In recent years, *combined-cycle* power-generation systems (Figure 3) have been developed to achieve more efficient fuel use with combustion turbines. In these systems, fuel is burned in a combustion turbine coupled to a generator, and the exhaust combustion gases are used to heat a boiler, producing steam to drive a steam turbine for additional generator power. Combined-cycle systems can increase the efficiency of fuel use up to 20% over conventional steam power generation.

In even the most efficient conventional steam power-generation systems, about two thirds of the heat units transferred to the steam from the original fuel are ultimately released, unused, in the low-pressure exhaust steam from the turbine. The major portion of this rejected heat is that which was required to vaporize the original boiler feed water into steam. This heat of vaporization is released by the steam when it is condensed back to liquid. However, the temperature at which this occurs is usually so low that little work can be obtained for further electricity generation.

While steam power plants are discarding the bulk of the heat content of the fuel they use, U.S. buildings and industries are independently burning fuels just to produce low temperature steam for space heating and process uses. Thus, combining these two operations by "cogenerating" electricity and useful low temperature steam or hot water can greatly increase the overall efficiency of fuel use.

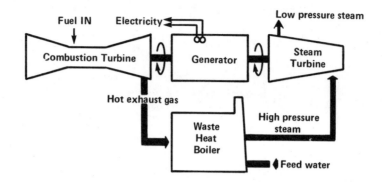

FIGURE 3.   Combined-cycle system for power generation.

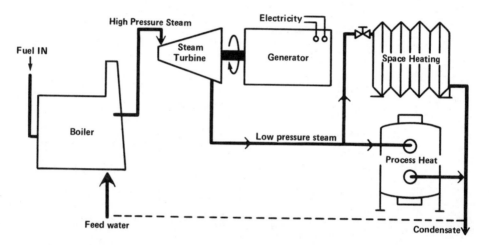

FIGURE 4.   Central utility cogeneration.

III. COGENERATION SYSTEMS

The dual purpose use of steam for electrical power generation and heating is not new. Utility central station steam/electricity cogeneration and cogeneration within industrial plants have been practiced to some extent for more than 50 years. However, some concepts such as the "Total Energy System" are a relatively new approach to cogeneration and are yet to be fully evaluated.

A. Central Station Steam and Electrical Power

Cogeneration in a large central station power plant (Figure 4) can provide district-wide space and water heating for commercial and residential buildings, low-temperature process heat for industry, or both. Together, space heating, water heating, and process steam account for about 30% of annual U.S. energy use (Figure 5). When electricity generation is included, more than 50% of the national energy demand consists of uses which might be served by cogeneration.

"District heating" using steam pipelines has been provided for many years by utility companies in central New York, Boston, St. Louis, and other U.S. cities. At one time, these systems cogenerated steam *and* electricity. In recent years, the utilities have preferred to supply district heating with steam from smaller, single-purpose, low-pressure boilers, and to produce electricity in large, high-pressure, generating stations. Potential energy savings from cogeneration have been insufficient to offset the capital, operat-

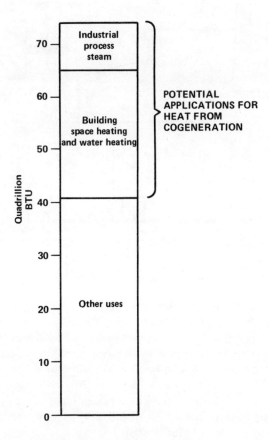

FIGURE 5.   U.S. energy use, 1976.

ing, and maintenance costs of the more complex dual-purpose systems. However as fuel costs increase, the economic benefits of central station cogeneration will also increase.

In Sweden, where energy costs have been historically higher than in the U.S., district heating systems are economical and currently supply about 20% of the energy required for space heating and hot water (Figure 6). In the Swedish systems, low temperature "reject" steam from turbines is piped through heat exchangers to heat water which is circulated to points of use. For applications such as space heating which require only relatively low (less than 200°F) temperatures, hot water systems are more economical than steam distributing systems which require more expensive piping and greater maintenance attention.

Though central station cogeneration to provide industrial steam was often attractive in earlier years (before 1950), few new systems have been installed. Currently, there are several locations in the U.S. with the required concentration of industrial steam demand. However, the economic, regulatory, and institutional problems of a combined venture have usually deterred new cogeneration programs.

This form of cogeneration is discussed more fully in Chapters 2, 4, and 10.

B. Cogeneration Within Industrial Plants

In the early 1900s, real (deflated) energy costs were even higher than today. The larger industrial plants not only generated essentially all of their own electricity, but

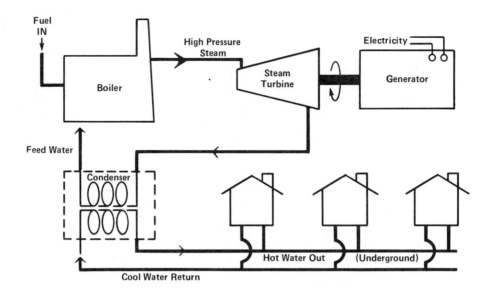

FIGURE 6.   European cogeneration system.

most of their electricity and steam was cogenerated. Although industry today purchases most of its electricity from utilities, about 100 billion kWh, one seventh of the
annual industrial electricity of the U.S. is still generated by the industrial user (Figure
7). About one half of this electricity comes from cogeneration systems.

For many plants in a few major industries like aluminum, paper, chemicals, and
petroleum (which use large amounts of both electricity and steam), energy economics
have consistently favored cogeneration. These plants have not only maintained cogeneration systems, but have continued to install new or more efficient dual-purpose
systems. One company in the chemical industry estimates that their modern cogeneration system involving combined cycle operation (Figure 8) is 30% more fuel efficient
than their cogeneration system of 30 years ago.

About 50 billion kWh of industrial electricity are generated annually from otherwise
"waste" energy, saving 250 trillion Btu per year, or the equivalent of 120,000 barrels
of oil per day.

These systems are discussed in greater detail in Chapters 3 and 11.

C. Total Energy Systems

Recently, the Department of Energy (DOE) has been looking at smaller, dual-purpose systems that could provide both electricity and heat to the concentrations of users
in large apartment building complexes, shopping centers, universities, or industrial
parks. Some of these proposed systems have included refuse processing to recover the
fuel value of the waste materials from the user community. In general, these smaller
systems suffer from high costs. Capital costs per unit of steam or electrical output for
small boilers and generators can be 50 to 100% greater than those for large industrial
or central station units. High maintenance and labor costs per unit of output make
these systems expensive to operate. User demand for electricity and heat in residential,
commercial, and industrial centers can vary hourly, daily, and seasonally, sometimes
as much as fourfold. Thus, while much of the equipment might stand idle during slack
periods, standby equipment may be needed to ensure reliability in peak-demand periods. Underused or redundant equipment adds to the per unit cost of cogenerated
electricity and heat.

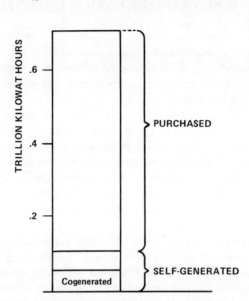

FIGURE 7.   Industrial electricity use.

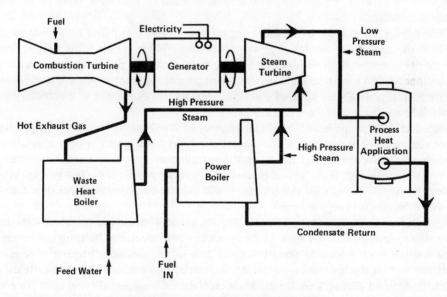

FIGURE 8.   Industrial cogeneration system.

With technological advances, small cogeneration applications are becoming more economical. A least one manufacturer is introducing a small, integrated, gas turbine/boiler/steam turbine unit with high reliability (and, therefore, low operating and maintenance costs) for these applications. Also, DOE is supporting work on new technology, like fuel cells and fluidized bed combustion, that can make on-site, dual-purpose systems cleaner, more efficient, and therefore, more practical for residential, commercial, and small industrial centers.

Further details concerning total energy/MIUS systems are covered in Chapters 5, 12, and 16.

IV. ENERGY CONSERVATION POTENTIAL

The potential for conserving fuel through cogeneration depends on the users demand for electricity and heat, and also on the type of cogeneration system to be used. Steam turbine, combustion turbine, and diesel engine systems will each produce a different ratio of electricity to heat output. Fuel conservation will depend on a cogeneration system's ability to match the users overall needs for each type of energy output and to cope with fluctuations in the individual demands. Steam is the primary form in which the heat output from a cogeneration system can be utilized. Steam is widely used in industry, and to a somewhat lesser extent in buildings, to transport heat to a point of use. Almost one half of all the industrial fuel burned is consumed in generating versatile, easy-to-use steam. A large, well-designed, industrial boiler will convert 85 to 90% of the combustion heat of the fuel into steam (Figure 9). If this steam is used for heating, and if, after such use, the hot condensate water is returned to the boiler, almost all of the heat content of the steam can be effectively used. However, when steam is employed to generate electricity in a modern, high-pressure, high-temperature, single-purpose plant, only 37 to 40% of the fuel's combustion heat is converted to electricity. Most of the remaining 60 to 63% is carried off by cooling when the exhaust steam passes through condensers at the end of the electricity generating process (Figure 10). While "rejecting" this large (45 to 48%) proportion of a fuel's heat energy seems inefficient, the electricity produced is "high quality" energy with the capacity to do more work or generate higher temperatures than could be obtained from the rejected low pressure steam.

While cogeneration systems improve the overall efficiency of fuel use, some of the gain is at the expense of electrical power generation. When steam is discharged from a turbine at pressures high enough for heating uses, electrical power output is reduced proportionately. In a cogeneration system producing by-product steam at a useful pressure of 50 psi, only about 15% of the input energy will come out as electricity, and almost 70% will be in the steam (Figure 11).

Cogeneration systems based on diesel engines or combustion turbines can produce higher electricity to steam ratios. However, the ratio for a given system is relatively constant and must match the needs of the installation or it will restrict the energy conservation potential. If the actual demand ratio varies more than a few percent from the design (expected) ratio for the system, it will usually be necessary to simply discard and waste the excess energy output.

Cogeneration systems may encounter highly variable demands for electricity and steam when serving end uses such as mechanical work, electrical lighting, or heating. Space heating load is high in the winter and low in the summer. Electricity requirements are highest during the evenings. An industrial plant may have a peak steam or electricity demand during a particular shift each day. Consequently, demand for electricity may not coincide with a corresponding demand for the cogenerated heat. A useful, fuel-conserving cogeneration system must be able to produce electricity and heat (e.g., steam) in proportions to match the "load" or array of uses served *whenever* required. In contrast, a prime attraction of conventional, single-purpose systems for process steam, space heating, or electricity generation is that the fuel input and the operating characteristics of the system are controlled by only one output demand.

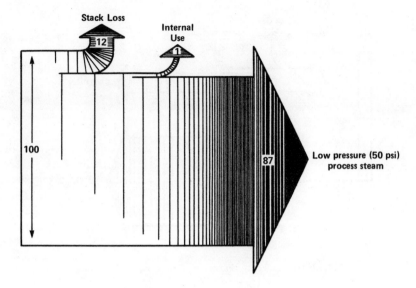

FIGURE 9.   Boiler energy flow.

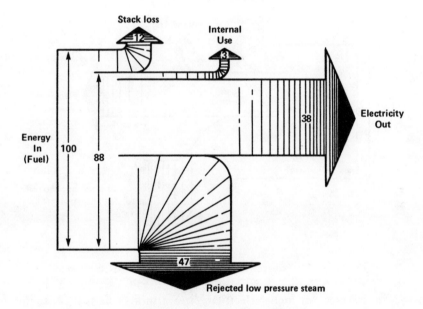

FIGURE 10.   Power plant energy flow.

Different modes of cogeneration will vary in their output ratio of electricity, high-temperature, and low-temperature heat. A basic steam turbine cogeneration system produces only about 15 Btu of electricity with each 70 Btu of steam (Figure 11). A combined cycle cogeneration system using a combustion turbine and an exhaust-heat boiler produces 30 Btu of electricity with 47 Btu steam load (Figure 12).

A diesel cogeneration system using an exhaust-heat boiler produces 36 Btu of electricity with each 27 Btu of steam load (Figure 13).

These ratios of electricity to steam are based on a steam pressure of 50 psi, which is somewhat low for average industrial use. For the higher pressures required in many steam systems, the electricity to steam ratio will be lower in the case of steam turbine cogeneration, but higher for the combustion turbine or diesel engine modes. However,

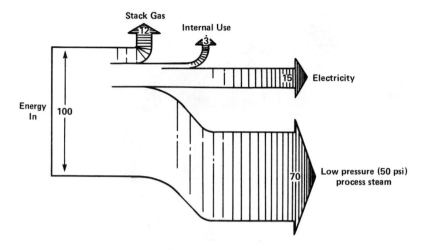

FIGURE 11.   Cogeneration energy flow, steam turbine system.

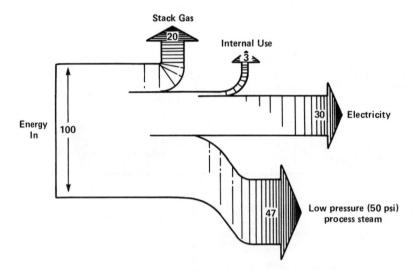

FIGURE 12.   Cogeneration energy flow, combined-cycle system.

in any system designed for higher electricity/steam ratios, the fuel consumption per unit of electricity also increases and the energy conservation potential is diminished. The overall efficiency of fuel use in a cogeneration system will, therefore, depend upon the type of system installed as well as the user requirements for electricity and high- or low-pressure steam from the system.

Cogeneration is currently favored where an industrial plant can satisfy its needs for electricity and steam with a simple, basic boiler and steam turbine system. Such plants will use 65 to 80% of their energy demand as low pressure (25 to 150 psi) steam, and the remaining 20 to 35% as electricity. For example, these conditions are found in many paper plants, and the pulp and paper industry is a leader in industrial cogeneration. Approximately 30% of the electricity used in the pulp and paper industry is co-generated with the steam production in paper mills.

In general, industrial cogeneration systems are designed primarily to satisfy the heating requirements of an installation, and electricity is considered only a by-product.

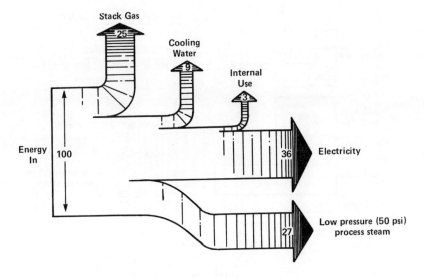

FIGURE 13.   Cogeneration energy flow, diesel system.

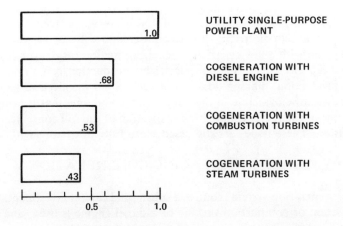

FIGURE 14.   Energy requirements per unit of electricity.

This electricity might be sold to the local utility grid or used internally. In either case, it replaces electricity conventionally generated in single-purpose power plants. Cogenerated electricity, depending on the system used, is produced with only 40 to 70% of the fuel normally required in an efficient, single-purpose power plant (Figure 14).

The potential for fuel conservation thus depends upon how much by-product electricity can be cogenerated, which in turn depends upon both the steam or hot water heating demand of the user and the mode of cogeneration employed. For a given heating-energy output requirement, steam turbine cogeneration will yield by-product electricity with a maximum fuel saving equivalent to 30% of the energy in the steam output. In combustion turbine and diesel engine systems, the overall fuel savings can be equivalent to 70 and 85% of the respective steam-energy output (Figure 15). However, because these latter two modes produce such high ratios of electricity to steam, the fuel saving *per unit of electricity* generated is actually less than for the steam turbine mode.

The overall potential for fuel conservation using cogeneration depends upon the relationship between the size and characteristics of user demands for low-temperature

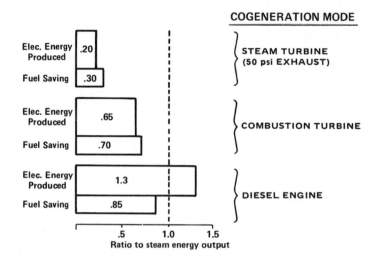

FIGURE 15.   Ranges of conversion efficiency.

heat and for electricity and the mode of cogeneration employed. It is unlikely that cogeneration systems which fulfill the operating and economic requirements of an individual user will also produce maximum oil and gas savings from a national viewpoint. A user with a high electricity/low steam demand will not be well served by a cogeneration system with a low electricity/steam production ratio, even if that system provides fuel-cheap electricity. Though electricity production can "ride" industrial process steam production, making dual use of fuel, either the electricity or the steam cogenerated may not be available in the amount or at the time that it is needed. This conflict is but one of the factors hampering realization of the full conservation benefits from cogeneration. These issues are discussed more fully in Chapters 10, 11, and 12.

V. CHALLENGES FACING COGENERATION

Installing a cogeneration system requires a substantial capital investment. Depending on whether a steam or combustion turbine or a diesel engine is used, and the size of the system, initial cost will range from $250 to $2000/kW of electrical capacity. Thus, a small 500-home district heating system or an intermediate-sized industrial plant may require more than a million-dollar investment. Although, to stretch national fuel supplies, it would be desirable to make dual use of fuel by cogenerating electricity and steam wherever one is now being produced alone, local economics limits cogeneration applications. Investments in cogeneration facilities will be withheld where industry, utility, and community decision makers find investment, maintenance, and fuel costs are not balanced by the expected annual fuel savings. The sharp increase in energy costs over the past few years should make cogeneration investments more attractive. However, other considerations create continued uncertainty about the cogeneration economic benefits. Some are presented below.

A. Consumer Willingness to Support New Heating Systems

Large-scale public sale of by-product steam or hot water by central utility stations will require acceptance by potential customers to justify the cost of installing piping for heat distribution. However, until such systems are installed and operating, potential customers will be concerned about reliability, future costs, adaptability to their individual needs, foreclosure of future alternatives, and aesthetic features (e.g., the size, location, and appearance of distribution facilities). In many cases, the on-site heating systems now in use by potential customers will need to be replaced or revised, entailing duplicate consumer expenditures. Such concerns may make it difficult to obtain commitments to using cogenerated steam or hot water systems. Further consideration of this factor is found in Chapter 16.

B. Cost and Availability of Fuels and Electricity

The greatest electricity production and overall fuel conservation are possible when combustion turbines or diesel engines, rather than steam turbines, are used in a cogeneration system to supply a given heating demand. However, present combustion turbines and diesel engines require natural gas or distillate oil fuels, both of which are becoming more costly and less available. As the prices of gas and petroleum increase, cogenerated electricity from these fuels may not be able to compete economically with electricity from large coal-fired or nuclear utility plants. Although research and development is proceeding on combustion turbines capable of burning alternate fuels, steam turbines fueled by coal, combustible waste, uranium, or even wood may be the cheapest mode of cogeneration.

The value of cogenerated electricity is uncertain. In installations where all of the cogenerated electricity can be used internally to displace purchased electricity, the value will equal the cost of the replaced electricity. However, any auxiliary electrical service ("standby" service) required to meet demands that the cogeneration system cannot handle may have a premium cost (see Chapter 13). If the cost of such auxiliary service is too high, it can significantly decrease the value of the cogenerated electricity.

If maximum electricity were to be cogenerated with the total U.S. industrial steam base today, less than 25% of that electricity could be utilized by the systems that generated it. The excess electricity would have to be purchased by the local utility company or transported to other potential users. The value of that electricity would depend on the quantity available, when and how long it is available, transmission costs, and the specific needs of the purchasing utility or user. These considerations are discussed further in Chapter 13.

C. Environmental Constraints

Direct coal-fired steam boilers combined with steam-turbine operated generators are the oldest, most well know and accepted cogeneration systems. However, coal systems are subject to new environmental protection requirements restricting the sulfur oxide, ash particles, and other pollutants in the flue gases. Existing and proposed regulations for this purpose have been both ambiguous (e.g., requiring the use of undefined "Best Available Control Technology") and increasingly restrictive. These constraints tend to decrease the attractiveness of coal-fired cogeneration facilities.

D. Regulatory Constraints

Industrial cogeneration systems which would sell excess electricity to a utility might be subject to public utility regulatory procedures. Administered at the state level, such regulations and requirements vary significantly from one location to another and pose a challenge to unified national energy policy planning (see Chapter 13).

E. Institutional Constraints

The reliable operation of a cogeneration system requires skilled personnel. These people are not readily available, and a new operator of a cogeneration system will have the task of training employees to operate the installation. Energy sytems carry the vital "lifeblood" of an industrial plant. Steam or electrical power failures are among the most serious problems that can occur. When both the steam system and the power system are integrated through cogeneration, a system failure becomes even more serious. Some industrial managers have viewed cogeneration as "too many eggs in one basket."

Cogeneration Methodology

SECTION 1

COGENERATION METHODOLOGY

PREFACE

The four chapters in this section comprise a broad look at the major forms of cogeneration and will acquaint the reader with the overall scope of dual-purpose energy systems.

Chapter 2 gives specific attention to the production and sale of cogenerated industrial steam as a by-product to normal electricity generation from a large, central utility power plant. Chapter 3 considers industrial cogeneration plants dedicated to a single user. For large installations, these would be owned and operated by the user. For small installations, third-party operation is described. Chapter 4 discusses district heating systems wherein the central utility supplies steam to multiple customers within a service area, in a manner analogous to their electrical power supply. Chapter 5 has to do with smaller integrated cogeneration systems designed to provide both heat and electricity to an institution (e.g., university) or a concentration of users such as a large multifamily housing development.

Each cogeneration methodology is discussed in terms of historical trends, prime application areas, technological alternatives, and energy conservation aspects. To a large extent, the various concepts are illustrated by describing specific examples of operating installations.

Chapter 2

COGENERATION OF INDUSTRIAL STEAM AND POWER IN CENTRAL UTILITY STATIONS

Charles D. Glass

TABLE OF CONTENTS

I. INTRODUCTION

Cogeneration is not new to American utilities. It has been practiced in this country for several decades. While some steam-electric plants are owned and operated by industrial concerns for their own use, many others are owned by utilities who sell steam and power to adjacent industries. Several of these plants are located in or adjacent to oil refineries and chemical plants where the balance between process steam loads and electric requirements make the economics attractive for cogeneration. Much of the steam and electric power used by the Exxon Bayway Refinery in New Jersey has been supplied by the cogeneration plant owned and operated by Public Service Electric and Gas of New Jersey, which was constructed more than twenty years ago. DuPont has been served at their Chambers Works in New Jersey by a cogeneration plant owned and operated by Atlantic City Electric since 1928. Gulf States Utilities built a cogeneration plant in 1929 to serve the steam and electric power needs of Exxon and the Ethyl Corporation at Baton Rouge, Louisiana. This latter operation, which has worked to the satisfaction of both the utility and the industrial customers for fifty years, is described below.

The Gulf States Utilities plant at Baton Rouge was originally conceived to meet the needs of both Exxon's petroleum refinery and Ethyl's tetraethyl lead plant. The refinery made heavy use of steam while the tetraethyl lead process was an intensive user of electric power. The energy usage of the customers has shifted somewhat over the years of operation, but the plant is still capable of supplying about three million lb/hr of process steam. The predominant amount of the steam supplied to process is delivered at 150 psig, but a portion of it is supplied at 600 psig.

The plant has an enviable record of reliability. Only twice in its history has the plant been shut down completely — once when the fuel gas supply was disrupted, and again when lightning struck a utility pole outside the plant perimeter and traveled into the main switching room in the plant. In both instances, service was restored in an acceptable period of time. The reliability is basically due to the plant design, including the large number of intermediate-size boiler and turbogenerator units that are in use. Three of the steam turbine units can be operated either in the condensing or the extraction mode. This enables the plant to respond to the swings in the steam take by the customers.

Figure 1 shows a layout of this plant.

II. TRENDS IN CENTRAL UTILITY COGENERATION

In the past two or three decades (when the majority of our rapid industrial expansion took place), the most economical way to produce process steam was with simple gas-fired boilers operating at pressures too low for efficient electric power generation. Power needs were primarily purchased from utilities, which built larger and larger plants producing power at costs that continually declined until the latter part of the 1960s. Industrial plant design and energy requirements were based on the economics that prevailed at the time. The dominant factor was the abundant supply of inexpensive natural gas. Natural gas was a waste product when it came into use as a boiler fuel. Only in recent years has it become evident that the highest value-use of natural gas was for petrochemical feedstocks and special heating applications. Even after more useful purposes were discovered for natural gas, regulation at low prices encouraged the continued use of natural gas as a boiler fuel. It was not until the early part of the 1970s that the price of gas began to reflect its true importance.

As the price of energy increased, industrial companies began to reexamine the eco-

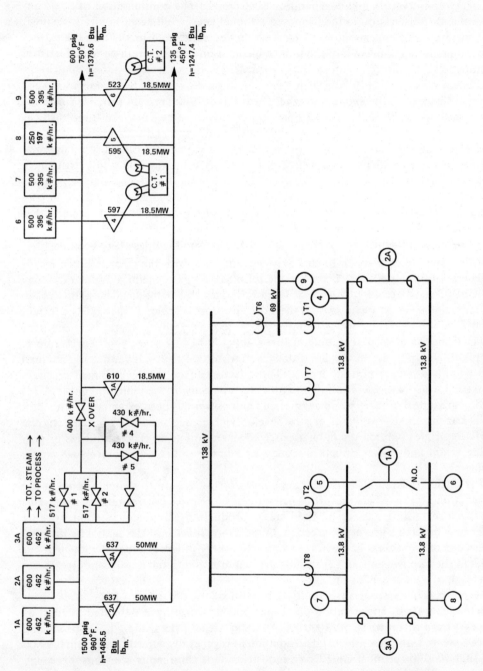

FIGURE 1. Gulf States/Exxon/Ethyl layout C.T. = condensing turbine and T = transformer.

nomics of their energy usage. Capital investments in energy conservation became economical. Many heat recovery ideas that had not been good investments prior to this time now became justifiable.

When the increased cost of energy was coupled with the embargo on oil, the interest in energy conservation and cogeneration accelerated. The consideration of coal conversion and cogeneration in the proposed national energy bill has also kindled a great deal of interest in cogeneration projects. A survey conducted in 1977 by the Edison Electric Institute indicated more than 40 cogeneration projects were under some form of study.

Some of these projects involve capturing the waste heat from process facilities, such as a petroleum refinery coking unit where heat recovery boilers can produce steam for power generation. Other projects are being considered where gas- or oil-fired turbines with exhaust-heat boilers would be used to cogenerate steam and electric power.

The existing rules of regulatory bodies, together with the probability of national legislation that would mandate switching from gas for industrial boiler fuel, have caused numerous studies to be undertaken. The probable shift to coal logically causes potential users to take a fresh look at cogeneration. The economies of scale in coal operation make it more economical to build central steam facilities for a plant or a group of plants in a given area. The economics for including at least some cogeneration are rather clear cut where new large-scale coal-fired steam facilities are to be built.

If the expected conversion to coal proceeds, the area with the greatest potential for cogeneration in this country is among the refining and petrochemical industries along the Gulf Coast. Here are clusters of large refineries and petrochemical plants which use large quantities of steam around the clock for process heat and have fairly constant electric power requirements. A number of these clusters have aggregate steam and power demands of over four million lb/hr and 400 MW, respectively. Furthermore, the plants within these clusters are usually well within a 5 to 6 mile radius, the nominal upper limit for steam pipeline length. These factors favor the development of large, centrally located, cogeneration plants to serve the needs of the industrial clusters. Indeed, a great deal of money and engineering manpower has been expended by the local industries in recent years in efforts to develop viable cogeneration projects along the Gulf Coast. Not only do the industries see cogeneration as an efficient means of providing steam and power, but it also may ease uncertainties over future fuel supply problems.

It is difficult to establish an economic basis for a decision to switch to coal at the present time, even when the efficiencies of cogeneration are included. Often, this would mean retiring gas- or oil-fired steam production equipment in good working order long before its useful life is ended. Quite likely, this transition cannot be justified unless one of two things happen: (1) a national energy policy is adopted with clear-cut rules on fuels conversion, or (2) an industry requires additional new steam production as a result of a plant expansion.

Recent feasibility studies (presently unpublished) for operating a large, newly constructed, coal-fired, cogeneration plant on the Gulf Coast have estimated the cost of process steam at up to $6.50/1000 lb. At today's fuel prices, the cost of producing process steam with presently installed equipment is generally agreed to be around $2.00 to $3.00/1000 lb for most Gulf Coast industries. One can readily see from this rough cost comparison why actual conversion is moving slowly.

There are many uncertainties involved in a wide range of factors that affect the decision making process on fuel conversion and cogeneration:

1.   Future availability of the various fuels

2.   Government policy affecting fuels pricing
3.   Possible use taxes on the burning of oil and gas
4.   Tax incentives for cogeneration
5.   Confusion over regulatory rule making
6.   Environmental constraints

The environmental permitting process represents a major deterrent to efforts to organize large coal-fired cogeneration projects. While there are indications that national energy policy will encourage cogeneration and the burning of coal, the laws governing environmental matters will act to prohibit the burning of coal in many locations. It is estimated that the typical time for securing environmental permits is from 24 to 36 months. This results in extending the lead time for constructing new cogeneration plants to 5 to 7 years. The lead time for constructing conventional new industrial plants is 24 to 36 months. This difference in planning horizons between the cogeneration plant and the industrial facility causes serious difficulty in bringing a project to fruition.

Where the negative factors and uncertainties can be overcome, and a decision can be made for constructing new steam-producing facilities that incorporate cogeneration, there are reasons both from the industrial and the utility viewpoint to have the utility active in the project. Among the utility benefits from cogeneration are

1.   Stabilize industrial load
2.   Source of new generation
3.   New business opportunity

It is difficult to make general statements regarding cogeneration benefits that will apply to utilities on a broad basis because business conditions vary so widely among the companies. The industrial business of some companies is small in comparison to the residential and commercial business, and characteristics of the industrial customers may provide little opportunity for cogeneration. In addition, there are wide variations in system load factors as well as differences in the fuels being utilized. However, in areas of the country where industrial customers have substantial needs for steam and electric power on a continuous basis, cogeneration may be desirable. To be attractive to a central utility plant, total steam loads would likely have to exceed one million lb/ hr. The areas where refining and petrochemical plants exist have already been pointed out as prime market areas for cogeneration.

The utilities serving these areas have large investments in generation plants for supplying the industrials. If the industrials were to relocate as a result of a lack of energy, or if they opted to supply their own needs, the utility could be adversely affected. Therefore, these utilities should have a natural interest in being involved in cogeneration projects to stabilize the business in which they have already invested to serve.

Properly coordinated with the overall generation planning for a utility, new cogeneration plants may offer the most economical source of new capacity. Even though the cogeneration plant may generate only a portion of the electrical requirements of an industrial complex, generation capacity would become available for the use of other classes of customers.

An example of this is the Consumers Power Company nuclear plant under construction at Midland, Michigan. This plant will supply both the steam and electricity needs of the Dow Chemical Company plant at Midland, while producing the bulk of its power output for general distribution.

Cogeneration plants also offer a utility the opportunity to enter into new business ventures (see Chapter 3). It may provide a utility with a higher return on investment opportunity.

There are benefits to the industrial plant from utility participation in a cogeneration venture:

1. Third party to organize and operate plant
2. Expertise for operating and maintenance
3. Buffer for regulation
4. Fuels management
5. Minimize reserve requirements
6. Financing

In attempting to organize a group of industrials to support a central cogeneration project, one of the most difficult things to overcome is the lack of uniformity among the participants on a wide range of issues. For example, the group will not all have the same creditworthiness. Also, problems will arise from varying degrees of ability to tolerate steam and power interruptions. Such issues must be resolved into positions that all can agree upon. The utility may play a useful role as a third party acting as a catalyst to promote agreement.

Many industrial companies are not staffed to operate power generation, having concentrated their efforts strictly on manufacturing their own product line. The utility may be in a better position to provide a capable group of people to operate and maintain the plant. The same would be true for fuels management* since the utility is already a large purchaser of fuels in its normal business operation.

At the present time, the regulations governing the generation of electricity would make the participants in a cogeneration project subject to the Federal Power Act and the Utility Holding Company Act if one of them owns more than 10% of the electrical generating facilities. It is anticipated that the National Energy Bill will alleviate these regulatory constraints on cogeneration. However, until this bill actually becomes law, all planning must be done with the existing statutes in mind (see Chapter 13). This factor is the primary reason several projects that are presently in planning will have the utility own the turbine generators even though the industrial partners may own the boilers or steam plant.

Examples of such projects are the Celanese Corporation plant at Pompa, Texas where a 30 MW generator owned by Southwestern Public Service Company will cogenerate electricity, and the CAM project in Texas City, Texas where the Community Public Service Company will own the generating unit and return exhaust steam to three industrial users.

The involvement of a utility may make it possible to achieve the lowest capital cost in financing a project. One of the basic reasons for this is that the utility can maintain a favorable credit rating (and interest cost) and a higher debt to equity ratio than industries. If the utility is able to finance the project, it may afford the industrial plant the opportunity to convert to a more favorable long-term energy source without adverse impact on their balance sheet.

When the utility owns at least the generating equipment in the project and operates it as an integral part of the utility system, the amount of backup power needed may be minimized. This is particularly true where all cogeneration power flows into the utility system and the industrials' electric power requirements are purchased under the

---

*   Developing and maintaining an assured, reliable supply of fossil fuel at minimum cost to a large steam plant is a demanding activity that is little understood or recognized by the general public.

regular published rate schedules. Although the sale of power from a utility to a self generator is not an uncommon situation, the question of back-up power for cogeneration projects has created a great deal of discussion (see Chapter 13). This question will eventually be settled on a nondiscriminatory basis that is fair both to the cogeneration project (regardless of whether it is owned by the utility or the industries) and to the other classes of customers of the utility. Backup power should be priced on the basis of the cost of providing that service and should take into account the allocation to peak load responsibility of reserve margins. Again, this is discussed more fully in Chapter 13.

Both the steam and electric power must be available on an uninterrupted around-the-clock basis for most industries. For this reason, the utility-operated cogeneration project serving this type of industry must be designed with multiple boilers and have the steam supply as a first priority. The number of boilers, turbines, and generators used must allow for equipment to be taken out of service for maintenance and must be designed to allow for emergency shut-down of equipment from time to time.

There are benefits from cogeneration to the general rate payers of a utility:

1. Favorable effect on power rates
2. Promotes economic development
3. Energy conservation

The participation in a viable cogeneration project by a utility will often provide power at a lower cost than other current alternatives, or alternatively, it may free up capital investment dedicated to the industrial sector. Therefore, there will be a reduction in the amount of generating capacity supplied by the general ratepayer through the normal utility construction budget. The beneficial effect should exist regardless of whether or not there is power exported from the project into the utility system.

In summary, there are substantive reasons for utilities to participate in industrial cogeneration projects. Industry receives benefits, the utilities' financial position should be enhanced, and for the regular customer of the utility, rates will be lower. The key to participation of the utility lies in creating some incentive for it to be involved in a program that may be considered outside its normal line of business. This incentive probably can best be achieved by exempting the project from regulation. This will allow the return from the investment in the project to be a matter of contract negotiations between the utility and the industrial customers and not dictated by a regulatory body.

Chapter 3

COGENERATION IN INDUSTRIAL PLANTS

R. W. Barnes and Paul Hodiak

TABLE OF CONTENTS

I. BACKGROUND

The dual purpose use of steam in industrial plants has been practiced from the earliest days of industrial steam use. It is interesting to note that today, generation of power as a by-product from process steam production is receiving increasing attention, whereas in an earlier day the opposite consideration was paramount. In the late 19th century, industrial steam was produced primarily to operate steam engines for power (mechanical and electrical), and the innovation was using the exhaust steam for heating purposes. Figure 1 is a diagram, published in 1895, of an industrial cogeneration system in that day.

Such an early system would have had a steam-supply pressure of 60 to 75 pounds per square inch (psig) with exhaust at slightly more than atmospheric pressure. Under these conditions, the heat rate for the cogenerated power was about 14,000 British thermal units (Btu)/kilowatt hour (kWh). This might be compared with the 4500 Btu/kWh heat rate attainable from present-day industrial cogeneration systems.

Although industrial steam was initially used primarily for mechanical power, the early 20th century brought rapid electrification of many industrial power needs. In the 8 years from 1912 to 1920, industrial electricity use almost tripled, growing at an average rate of 14%/ year. With this conversion came the installation of electrical generators and the industrial "boilerhouse" became the "powerhouse". The self-generation of industrial electricity was the general rule at this time. From the beginning, however, economies of scale were significant in electrical power generation, and central-station utility power began to supply an increasing portion of the industrial requirement. Self-generation of electricity by industry has declined from essentially 100% of industrial electrical use in 1910 to about 10% today (Figure 2). In terms of electricity output, however, industrial generation grew from 17 GWh (gigawatt hours) in 1920 to 110 GWh in 1969, and has since been decreasing.

Not all self-generation of power in industry is cogeneration. In fact, it appears that only about one half[1] or less of the electrical power generated by industry is from dual-purpose steam (or heat) production.

Two primary reasons account for the decline in attractiveness of industry cogeneration. In the post-World War II era, industrial electricity purchase costs declined significantly (on a constant dollar basis) by taking advantage of cheap natural-gas fuel and the economies of scale from increasingly larger central utility generating plants. Also, in the 1950s, the industrial "package" boiler came into use. This boiler was shop-assembled and could be installed in an industrial plant much more quickly and at a much lower cost than a field-erected boiler. The package boilers, however, were designed only for gas (or clean oil) fuel and for steam pressures too low for efficient power generation. Under these conditions, the trend in U.S. industry was to install package boilers for process heat and to purchase electricity that would have formerly been self-generated. Where industrial power generation grew by 6.6%/year from 1950 to 1955, growth was only 1.1%/year from 1965 to 1970.[2]

The economics for industrial cogeneration are influenced primarily by energy costs, plant operating schedules, and the electricity/steam loads. Although the true feasibility of a given potential application can be ascertained only by an application-specific analysis, it is possible to develop some general guidelines for a rough economic screening. These guidelines are shown in diagram form in Figure 3. While these diagrams indicate a cogeneration feasibility region for each variable, there are other factors which may or may not be present in a specific case to enhance the feasibility. These include:

1.   Questionable service from the local utility to meet electrical load demand.

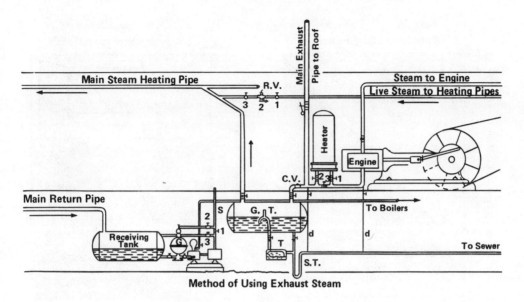

FIGURE 1. A nineteenth century industrial cogeneration system. (From *Eng. Rec.*, 1895. With permission.)

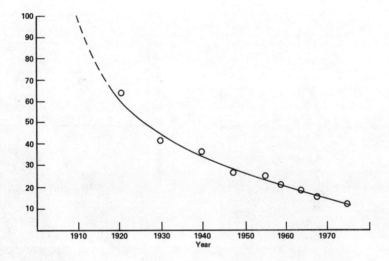

FIGURE 2. History of industrial self-generation of electricity. (Data from Historical Statistics, Colonial Times to 1970, and Census of Manufacturers: Annual Survey, 1975, U.S. Department of Commerce, Washington, D.C.)

2. Availability of waste fuel materials.
3. Obsolete boilers needing replacement.
4. Excess steam capacity in existing systems.

The 10% of industrial electricity that is self-generated is primarily concentrated in a relatively few industries which are listed in Table 1. The paper industry, as would be expected from the general feasibility criteria given above, is the largest practitioner of self-generation and cogeneration. Although exact data for the amount of cogeneration is unavailable, estimates solicited from paper industry representatives indicate that about 75% of the electrical power is produced in association with process steam. Esti-

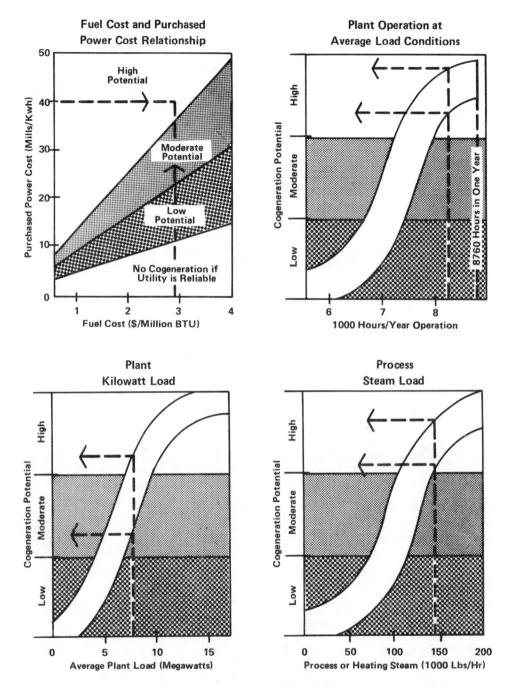

FIGURE 3.   Feasibility indicators for cogeneration potential. (Courtesy of W. B. Palmer, General Electric Company.)

mates of the level of cogeneration in the chemical, petroleum, and food processing industries were in the range of 75 to 80%. However, the primary metals industry has relatively little requirement for low pressure steam, and cogeneration was estimated to represent only 10% of self-generated electrical power. Overall, it is believed that about one half of the total electricity generated in industrial plants currently is produced in cogeneration systems.

TABLE 1

Self-Generation of Electricity by U.S. Industries:
1975

Industry	Electricity generated for own use (10^9 kWh)
Paper	24.4
Chemicals	15.0
Primary metals	13.8
Petroleum	4.2
Food processing	2.5
Other industries	3.4
Total	63.3

From **Anon.**, Annual Survey of Manufacturers
1975, Fuels and Electric Energy Consumed,
M75(AS)-4, U.S. Department of Commerce, Washington, D.C., 1975.

It should be noted that in the iron and steel industry (and elsewhere) a significant, but unknown, quantity of high pressure steam is used in turbine-drive compressors and blowers, and in many cases, the low pressure exhaust steam is used for process heat. Even though no electricity is produced, this represents an efficient dual-purpose use of energy and should be considered cogeneration.

II. TYPICAL CHARACTERISTICS OF INDUSTRIAL COGENERATION PLANTS

A. Large Industrial Plants

Two relatively new large-scale industrial cogeneration plants will be described. The first example typifies the use of coal in a basic steam-turbine system. The second exemplifies the use of gas (or oil) in a combined-cycle system.

FMC Corporation, Green River, Wyoming[2]

A new coal-fired steam plant was brought on stream in 1976. The new plant replaced existing gas- and oil-fired boilers and was installed to minimize the rapidly increasing fuel costs. The plant consists of two 650,000 lb/hr boilers with the old boilers kept in operable condition for backup. Steam is produced at the relatively mild conditions of 625 psig pressure and 750°F temperature required for compatibility with the old system.

Medium Btu, high-moisture, western coal is burned in a pulverized-coal combustion mode. The boilers are designed oversize to be able to maintain the required steam temperature and pressure when burning even subbituminous coals. The new boilers are also capable of burning oil or gas in combination with the coal. Thus, the plant has flexibility to cope with a variety of fuel supply conditions.

The boilers are equipped with control equipment to meet all presently foreseen air pollution restrictions. This equipment includes secondary combustion air injection (for NOX control), electrostatic precipitators, and circulating soda-ash-solution flue-gas scrubbers. It is estimated that the pollution control equipment consumes up to 5% of the total energy output from the plant.

The FMC plant has three 11.5 MW turbine/generators and three smaller units for a total of 43.5 MW. These units supply approximately 75% of the electricity required, with the remainder purchased from the local utility. The turbines are all extraction/ back pressure type and serve to reduce the steam pressure from the 625 psig at the boiler to the process steam pressure levels (200, 25, and 20 psig). Since the cogeneration operation supplies only 75% of the required electricity, consideration was given to installing steam-condensing turbine capacity to generate the additional electricity. This, however, was found to be economically uncompetitive with purchased power from the utility. Also, to minimize demand charges from the utility for standby power (when a cogenerating unit is down for servicing), multiple 11.5 MW units were installed rather than a single larger unit.

The entire steam/power system is monitored and controlled by a fully automatic computer system. This energy management system is patterned after those employed by full-scale utilities and is designed to keep the system operating efficiently and safely under almost any type of disruption or load change. The highly sophisticated control was justified economically on the expectation of fewer production interruptions, more efficient utilization of cogenerated steam and electricity, and reduced peak load demands for purchased power.

After more than three years of service, the new FMC cogeneration system is operating effectively and meeting the original design criteria.

2. The Dow Chemical Company, Sarnia, Ontario[3]

In 1972, Dow installed two 51,800 kW gas turbines exhausting through two multipressure waste-heat-recovery boilers. The new units were integrated into an existing cogeneration operation consisting of four high pressure boilers and two 30 MW extracting back-pressure turbine/generator sets. High pressure steam is produced in the boilers at 1400 psig and reduced across the steam turbines to 150 psig and 45 psig pressures for process use.

The gas turbine heat-recovery boiler units are highly efficient. They are equipped with supplementary firing facilities to utilize the excess oxygen in the turbine exhaust (which raises the gas temperature to meet superheat requirements). Final exhaust gases are reduced to a temperature of 135°F by generating low pressure steam (45 psig) and preheating feedwater for the conventional boilers. The gas turbine/heat-recovery boiler combination achieves a thermal efficiency of 86% and the heat rate for the electricity thus produced is less than 4000 Btu/kWh. Overall, the integrated power plant has a thermal conversion efficiency of 83%.

Unlike the FMC plant described previously, the Dow plant normally supplies 100% of their own electrical power requirements. The plant has the flexibility to vary the electricity to steam ("alpha") ratio from as low as 0.18 up to 0.55 and still maintain above 78% thermal conversion efficiency. They have an interconnection with the local utility whereby they can buy or sell power at a daily rate, whenever it is economically attractive to do so.

The cogeneration plant has computerized control to shed load when outages occur and to maintain normal operations at the most economical operating point.

B. SMALL INDUSTRIAL PLANTS

Applied Energy, Inc. (AEI) is a wholly-owned subsidiary of San Diego Gas and Electric Company. It was formed in 1968 to install a central steam plant for the Naval Training Center in San Diego. The central steam plant was designed to utilize a cogeneration system as the primary source of steam. The cogeneration plant was installed in 1971 and has been operational since that time. This plant led to two other Navy cogeneration installations. These are the Naval Station, San Diego, in 1976 and North Island Naval Air Station, Coronado, in 1977.

Each AEI plant has its unique design, but each has similar main systems. Electrical power is generated by gas turbines, and steam is produced in heat recovery boilers. To supplement the heat recovery boilers, a standby package boiler has been installed at each site. These package boilers have the capability of carrying the normal steam loads and serve as a backup steam supply. The package boilers are also used for periods of low steam demand when the operation of the heat recovery unit would not be economical. The normal steam outputs for these three military steam plants are in the range of 100,000 lb/hr.

The AEI business concept is somewhat unique. In these cogeneration plants, the local utility (San Diego Gas and Electric Company) owns the gas turbine and produces the electricity. They maintain and operate the turbine and can change the electric output at anytime there is a need for it in the electrical system. The operating and maintenance *costs* for the turbine are the responsibility of AEI as stated in a contractual agreement between AEI and the utility. The electrical output from the gas turbine is connected to the local grid and there is always a market for it. The electricity delivered to the grid is valued at the system-average generating costs and constitutes a credit against the operating and maintenance costs of the turbine. The contractual arrangement, actually, purchases waste heat from the turbine with a pricing mechanism that makes the electricity produced look like any other base-load supply of electricity. This, therefore, creates no hardship for local electric rate payers.

The exhaust-heat cost is related directly to the heat rate of the turbine. The higher the heat rate (less efficient), the more expensive the exhaust heat. This impact is somewhat mitigated by the ability of the higher exhaust-heat temperature (from a less efficient turbine) to produce more steam. The trade-off varies with the steam load pattern and the shape of the load swings as they affect the part load efficiencies of the turbine and the steam production. The general philosophy is to use the lowest heat rate possible and at times spill the excess exhaust heat where it remains economical. This is covered in more detail in Chapter 12.

The steam supply is provided under a contractual arrangement between AEI and the customer. The utility is *not* a party to the agreement. AEI deals only in the thermal energy of the steam. AEI sells the heat content of the steam, not the steam. The customer systems are all designed for condensate return.

The steam customer is contracting for continuous steam. This contractual arrangement makes AEI close to an operating arm of the client. Incentive for such an arrangement is the customer's desire to avoid making a capital investment for steam equipment and to avoid the problems of guaranteeing the fuel supply for his steam plant. There is one unquantified secondary motivation — the electric supply reliability presumably is increased with a turbine set "on-site".

Ultimately, AEI performs a broker function. It joins a willing utility with a willing customer through a third party (AEI) whose priorities look out for the interest of both the utility and customer. The tact required in preserving each party's priorities is the largest stumbling block to consummating a cogeneration deal.

The three Navy plants are not considered large industrial sites, but rather, moderate industrial loads. They also show cyclic variations and are less uniform than a conventional industrial process.

At the present time, AEI is in the final stages of constructing an unmanned, low pressure industrial steam plant for an industrial client in Chula Vista, Calif. This plant has an oil-fired turbine producing low pressure steam for a 7 day/week metal processing operation. The loads, unlike the Navy installations, are very uniform. Also, unlike the other three central steam plants, the customer at this installation has both water treatment and standby boiler capacity installed. The customer contracted for the cogenerated steam because it would be less expensive than self-made steam, considering

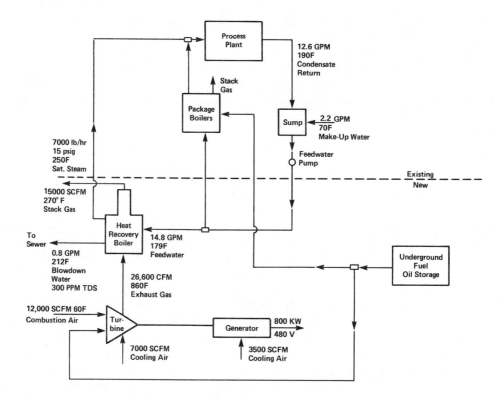

FIGURE 4.   Cogeneration system for Chula Vista plant.

fuel changes that he was being required to make. Mandatory end-use priorities would have forced a conversion to oil in place of propane for standby fuel.

This plant represents the smallest plant in size that AEI anticipates placing into operation. It is the largest plant that will meet California new-source air-quality regulations without serious economic trade-off penalties. Design details of the plant are discussed below.

As shown in Figure 4, saturated steam will be generated on a Deltak exhaust-heat-recovery boiler rated at 7000 lb/hr at 15 psig and 250°F. The exhaust heat is provided by a Solar combustion turbine set rated at 800 kW, 1000 kVA, 480 V, 3 phase. The combustion turbine is fired on #2 diesel stored on-site in three 12,000 gal underground fuel tanks. The fuel tanks are used for the utility turbine fuel as well as for the client's alternate fuel uses.

The existing gas-fired package boilers are used as a cold standby and for periods of high steam demand or during turbine maintenance. Unlike the large Navy plants, the industrial package boilers are being retained in operating order, thus reducing the AEI expense for a continuous steam supply. This means that the existing boiler feed water and chemical treatment can be used with no incremental cost. The same holds for the condensate return and steam distribution system.

The electric output of the generator is tied into the utility grid through a 480 V, 12.5 kV step-up transformer and its related switchgear. The electric output of the generator is a function of steam demand. This unmanned facility has controllers which allow remote monitoring and permit the output to follow steam load. There also is a mode of operation which allows the turbine exhaust to bypass the boiler and the turbine to be placed in base-load operation, independent of the boiler.

An important feature of this plant is the emergency operating mode. If outside power is lost or seriously reduced to a critical portion of the customer facility (i.e., the computer center), the customer can switch the cogeneration turbine electric output to feed the critical load. The turbine is automatically separated from the grid, and power is delivered to an uninterrupted power supply (UPS) system. The critical load must be partially powered-down to maintain the load in the maximum 700 kW range, but the UPS system lengthens the permissible time for cut over.

Under the emergency operation mode, the electric load determines the amount of exhaust gas energy available to the boiler, and a set of dampers modulates the amount of exhaust gas that enters the boiler to match the steam demand. When the turbine operates as an electric peaking unit for delivery of electricity to the grid, it has this same operating mode.

The anticipated fuel savings is 23% over separate conventional generation of electricity and steam. The peak thermal efficiency is expected to be 79%. The well-defined steam loads support the expectation that the projected fuel savings can be achieved.

The plant is designed for unmanned operation. There are local alarms that will be available for customer employees to observe when they are in the area. Operating status and data are telemetered to a central utility operating area, which allows monitoring and control by generation-station personnel. A roving operator will perform an inspection tour daily. There are several shutdown features on the cogeneration plant. Some malfunctions will put the turbine in an idling conditon with exhaust through the bypass stack. The Heat Recovery Boiler is designed to withstand a full loss of feedwater, but as with all features of this type, one hopes that it is never tested.

REFERENCES

1. Resource Planning Associates, The Potential for Cogeneration Development in Six Major Industries by 1985, U.S. Department of Energy, Rep., Resource Planning Associates, Cambridge, Mass., 1977, 15.
2. Schwiger, R. G., Elliot, T. C., and O'Keefe, W., Plant design today: a new challenge, *Power*, 37, Nov., 1976.
3. Zanyck, J. P., On-Site Power Generation Can Save Energy, paper presented at the Energy Conservation Seminar, Canadian Chemical Producers Association, Toronto, Canada, October 1976.

Chapter 4

DISTRICT HEATING SYSTEMS

Thomas E. Root

TABLE OF CONTENTS

I. INTRODUCTION

District heating systems which employ cogeneration fall into two categories: (1) urban utility systems serving central business, and (2) residential districts and on-site total energy systems such as those used by universities, hospitals, malls, etc. Since the total energy system is discussed in the next chapter, this chapter will concentrate solely on the urban utility systems.

II. HISTORY

The first recorded successful application of district heating was in Lockport, N.Y. in 1877. In that year, Birdsill Holley organized the Holley Steam Combustion Company, and in October, connected the first 14 customers to the system. Within 10 years, 18 new district heating systems were in operation across the country, from New York City to Denver, Colo.[1]

It did not take long for the infant electric industry to discover that the exhaust steam from the reciprocating engines used to drive their generators could be sold for a profit. The utilities primary concern was recovery of the cost of the distribuion system. Thus, the reject steam was sold at a very low cost to the customer. Soon, many cities had steam companies offering district heating. However, these initial applications of cogeneration were short-lived. Condensing steam turbines began replacing the reciprocating engine as the driver for the electric generators. This meant no exhaust steam for the heating system, and the condenser water was not of a high enough temperature to be useful. Electric plants became larger as technology progressed, and they were located farther away from the residential concentrations and commercial district heating areas.

Transportation of even extraction steam, if it was available, became impractical. As the larger condensing electric plants were commissioned, the small, obsolete plants in town were retired. The loss of the low-cost exhaust steam forced district heating systems into a new economic situation, away from cogeneration. At the same time, the availability of natural gas was growing rapidly, and the price was dropping. The capital cost of steam boilers, along with fuel and operating expenses, had to be recovered through higher rates, or the company would go out of business. Many steam companies folded.[1] Many others supplied their steam customers from inexpensive, low-pressure, gas-fired boilers without cogeneration.

Today, there are more than 45 district heating companies operating in the U.S., not counting the 14 colleges and universities with major steam systems, or the numerous other institutions which have central station steam heating.[2] Of the more than 80 billion lb of steam sold in 1976 by district heating systems, approximately 35% was from cogeneration sources.[2] Cogeneration has been an important contribution for larger systems for more than a quarter of a century.

High energy costs, due to mostly imported fuels, have kept European countries very conscious of energy efficiency for many years. Industrial and institutional cogeneration has been practiced extensively for nearly 30 years, resulting in a number of highly developed systems. The best known and most highly developed of these systems is district heating.

District heating systems exist in almost all European countries. Though an old idea, large portions of existing systems were installed during the period of reconstruction following World War II. Since then, district heating has taken a strong hold in Northern Europe and has become a matter of policy in Eastern European countries.

Though the precise number of consumers served by district heating systems in Europe is not available, an approximation can be made by analyzing the heating capacities of existing and planned district heating systems. A rule of thumb is that the per capita consumption for space and water heat is 3 to 4 kW under peak conditions.[3]

Much of the existing district heating is provided by refuse incinerators, electric plants, and oil-fired boilers. For the future, most countries find it desirable to use low temperature heat from nuclear plants as heat sources, with Sweden, Finland, Denmark, France, Switzerland, and Germany having advanced plans to utilize this heat source.

The Eastern European countries have followed the lead of the Soviet Union in establishing a formal government position on the utilization of heat produced in conjunction with electric production. The Soviet Union is the world leader in implementing district heating with rejected-heat utilization. There, 60% of the district heating is supplied by heat-electric plants, and about one third of all electric plants operate in this dual mode.[4] This situation has been expedited by the government position that rejected heat produced by large power plants should be utilized.

Several conclusions can be drawn from the experience of the European district heating systems, which must be kept in mind when considering implementation in the U.S.

1. The annual per capita space-heating consumption in European cities, where district heating exists, is approximately one half that of comparable American cities.[5] This is a result of more efficient conversion in central plants, a smaller per capita living space via apartments, and substantially higher insulation standards.
2. In all Northern and Eastern European Countries, district heating has been made cost competitive with other heating methods. This competitiveness can be attributed to higher fuel costs in those countries than in the U.S., to the installation of major portions of these systems during the rebuilding following World War II which resulted in lower capital costs, and from the governmental regulation and financing of district heating.
3. Almost all of the existing and planned district heating systems use hot water as the transport medium, instead of steam. Hot water systems have the advantages of lower capital costs because of their low pressure and temperature, lower system losses, and greater practical transport distances.
4. Because of cost, environmental, and energy conservation benefits, Europeans project substantial increases in district heating in conjunction with the development of nuclear power. This will be achieved, however, through government regulations requiring 100% connection in areas served by district heating.

III. TYPES OF SYSTEMS

A comparison of the main characteristics of steam and hot water heating systems are shown in Table 1. Most steam systems operate at high pressure and temperatures on a single main, i.e., no condensate return. These systems require very little user equipment and can be retrofitted into buildings equipped with hot water systems via a surface heat exchanger. In addition, steam can be used in absorptive air conditioners which are relatively inefficient, and is advantageous for processes such as sterilizing surgical instruments and in cleaning establishments.

The hot water systems are made up of at least two mains, a feed and a return, with pumping stations to maintain the system pressure. Hot water cannot be easily retrofitted into a building which has a steam system. For tall buildings, the user generally must install a pumping station to lift the water to the upper stores since the main system pressure is inadequate.

At the power plant end, there is little difference between the steam and hot water systems. Both utilize a boiler, and either a back-pressure turbine or an extraction tur-

TABLE 1

Characteristics of Steam and Hot Water District Heating Systems

	Steam	Hot Water
Temperature (°F)	300—500	150—250
Pressure (psi)	150—500	100—150
Main pipes	1	2
Pumps	No	Yes
Steam traps	Yes	No
Metering	Difficult	Easy
User investment	Low	Moderate
Transport distance (max.)	3 Miles	70 Miles
Type of pipe	Steel	Steel, concrete, plastic tile, fiberglass, etc.
System losses	Moderate	Low

bine, with or without condensing. In some cases, a combustion turbine with a heat-recovery boiler can be used, but because of the limited size of these turbines, they are more applicable to Multiple Integrated Utility Systems (MIUS) or Integrated Community Energy Systems (ICES) (see Chapter 5). The hot water system utilizes a steam-to-water heat exchanger, while the steam system usually does not use a heat exchanger. One of the primary concerns in both systems has been the maintenance of an oxygen- and carbon dioxide-free state, to reduce corrosion. Steam systems have solved this control problem.

In designing a district heating system, retrofit of existing electric power plants should be given first consideration as heat sources. However, the applicability of an existing power plant to district heating will depend on its design, age, location, and fuel type. From a technical point of view, the most important factor is design.

Most power plants use a condensing turbine from which some extraction steam can be taken while maintaining adequate steam conditions in the turbine. However, the effect that the change in external flow has on upstream turbine flow and on blade torque limits must be carefully analyzed in determining the point and amount of extraction. Because the temperatures and pressures required for district heating are low, the best point for extraction from a second law of thermodynamics point of view is on the low pressure side of the turbine as shown in Figure 1A. If this is not possible, then extraction from the high pressure side of the turbine may be used (see Figure 1B). However, this will cause a greater loss of electrical production than low pressure extraction, and may require a pressure-temperature (P-T) reducing station if steam is the transport medium for the district heating system. Finally, if it is not possible to extract steam from the turbine, then steam can be extracted from the main steam line to the turbine, as shown in Figure 1C. Again, a P-T station may be required.

For new cogeneration plants, or existing plants in which turbines are to be replaced, a back-pressure turbine can be installed, (see Figure 1 D). This is the ideal choice, since the plant can be controlled by either the *district heating* or *electrical system,* and since this type of turbine results in the highest thermodynamic efficiency for cogeneration application.

The ability to follow load on either the electrical or district heating system is extremely important in assuring a high capacity factor for the plant. This differs radically from the retrofitted electric plant with a condensing turbine which is controlled by the requirements of the electrical system. In the condensing plant, the load on the district heating system must follow the load on the electrical system, or steam must be dumped. This greatly reduces the plant efficiency.

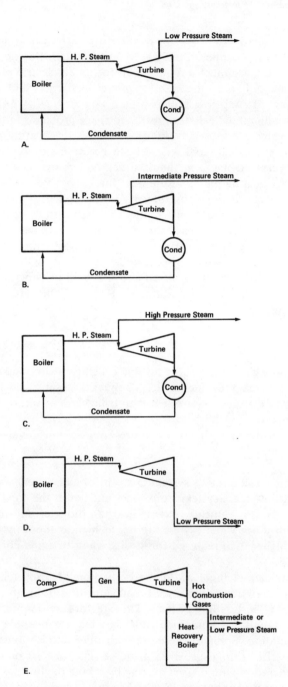

FIGURE 1. (A) Low pressure extraction from condensing turbine, (B) high pressure extraction from condensing turbine, (C) high pressure extraction from main steam line, (D) backpressure turbine, (E) combustion turbine with heat recovery boiler.

Another possibility is a heat recovery boiler on a combustion turbine, Figure 1 E. This arrangement causes almost no loss of electrical output from the turbine, but because these are considered peaking units on most electrical systems, they cannot serve as base-load units for district heating. A careful evaluation of the annual operating hours, fuel availability, and capital cost of conversion is necessary to determine if this arrangement is economically feasible.

Because of the relatively small size required for district heating boilers at this time (200 to 1000 MWth, as opposed to 3000 MWth for large electric plants),[5] the options available to a utility are much broader than for an all-electric plant. Coal-, oil-, and gas-fired package boilers in this size range are readily available and can be quickly installed. The use of coal-fired boilers presents an air quality problem, logistics (fuel delivery and handling, and ash disposal), and space problems in the urban and suburban locations which are most advantageous for district heating. However, if district heating is to be successful, coal will have to be the major fuel because of its lower overall cost. Recent advances in fluidized-bed boilers may make it possible to build coal-fired district heating plants in the near future without the use of scrubbers. Another option available is the refuse boiler. When applied to district heating, it offers the advantages of being located where most of the refuse is generated, thus reducing transportation costs of the refuse, and of burning a relatively clean fuel. However, because of the low density and variances in the composition of the fuel, refuse boilers are more commonly utilized as heat-only boilers.

Where possible, the boilers discussed above should be superheated. The use of superheated-steam boilers may be questioned. However, they are recommended for the following reasons:

- Capital costs for superheated boilers are similar to capital costs for saturated-steam boilers.
- There is more kW output at the turbine-generator per pound of steam input.
- Superheated-steam turbines are less complex than saturated-steam turbines.
- Superheated systems require less maintenance than saturated-steam systems of equal pressure.
- Superheated-steam turbines are more commercially available than saturated-steam turbines.

Whether the district heating system is steam or hot water, water quality must be kept very high. For steam systems, the water must meet the conditions for the boiler, which means very low chlorine, oxygen, carbon dioxide, and mineral content. The requirements for a hot water system are not as high as those for a steam system, but still are much higher than required for service water to avoid corrosion and sludge in the system.

The technical factors involved in the evaluation of district heating, as discussed above, can be broken down into three broad categories: (1) plant, (2) distribution piping systems, and (3) in-building systems. These factors are summarized in Table 2.

Many combinations of equipment and systems are possible for district heating through cogeneration. Some of these use innovative applications of technology, such as using the district heating system as a source-sink for heat pumps. Others combine industrial cogeneration with district heating by taking the low pressure steam from the industrial complex to supply the district heating system, or aquifers can be used to give high heat-load factors.

IV. DISTRICT COOLING

One of the key requirements for the successful implementation of district heating and cooling is the ability to supply as many services as possible on a year-round basis.

The addition of summer air conditioning capability to a district heating system extends the usefulness of the system, which results in better capital utilization, and thus, potentially higher profits to the utility and lower unit cost to the customer. Also, by being able to provide the total space-conditioning requirements for a building, there

TABLE 2

Technical Factors

Plant
  Type of boiler
    · Conventional
    · Fluidized bed
    · Heat recovery (CT)
  Type of fuel
    · Coal
    · Oil
    · Gas
    · Refuse
    · Wood
    · Nuclear
  Fuel Delivery and Storage
  Ash disposal
  System load and duration
  Boiler efficiency
  Heat exchangers (hot water)
  P-T stations
  Electrical system integration
  Retrofit alternatives
    · Extraction points
    · Back-pressure turbine
    · Boilers
  Siting and land use
  System losses
  System expansion alternatives
  Load factor — electric and heat
  Community impacts
Piping system
  Line losses
  System load
  Send-out temperature
  Return temperature
  System pressure
  Water quality
  Pumping stations
  Line size and expansion potential
In-Building System
  Enthalpy drop through buildings
  Building design
    · Thermal efficiency
    · Retrofit or new heat exchangers
  Number and type of heat exchangers
    · Heating
    · Hot tap water
    · Cooling
  Process Use

is a greater chance of public acceptance of district heating. Unfortunately, the countries which have extensive district heating have relatively mild summers, and thus, district cooling technology has not been developed to the same degree as district heating.

District cooling can be provided in two basic ways: (1) hot water or steam is provided to the users cooling equipment, or (2) chilled water is provided directly. Both of these means of cooling present a number of problems to the users or the utility.

Supplying hot water for cooling to the district heating system presents much the same situation as the steam case. However, higher temperature water must be supplied for district cooling than for district heating. This presents a potential problem for the utility and the user. If the system is operated at two temperatures, the hot tap water and heating equipment user must be sized for the lower winter temperature which will result in higher tap water temperatures during the cooling season. However, this technique allows greater electrical production from the power plant in the winter. Operating the system at the summer temperature year-round allows the use of smaller in-building hot tap water and space heating equipment, but the total electric generation from the cogeneration plant will be reduced because of the high winter operating temperatures for the system.

The other basic option for district cooling is to provide chilled water directly from the power plant to the community. This approach greatly reduces the in-building system costs. This is because the same heat exchange can be used for cooling and heating, and because it is potentially more energy efficient than supplying hot water or steam to inbuilding systems. However, the hot-tap-water heat exchanger must now be a separate unit. The greatest impediment to this approach is the requirement for a four-pipe system, i.e., two hot water lines and two chilled water lines. A dual piping system is necessary to supply both hot tap water and chilled water simultaneously.

V. RANGE OF APPLICATION

If buildings are considered separately by class, residential, commercial and industrial, then it can be shown that conventional single family homes cannot now be served economically by district heating. As building sizes increase through small apartment buildings, a gray area is encountered in which the economic viability is a function of system construction and operating costs, and building density. With high-rise apartments, approximately 100 units or more, there is a clear economic advantage for district heating from coal-fired plants as against individual gas- or oil-fired boilers.

Most commercial areas, such as central business districts, malls, and shopping centers, can be served by district heating, if they are not too far from the plant. Small, scattered businesses present the same density problem as single family homes. Included in the commercial class are such institutional buildings as government offices, schools, and hospitals. All of these can be excellent load for the system. However, some institutions, such as hospitals, have special requirements such as high reliability and certain process conditions, e.g., sterilization, that must be met.

Industrial applications of district heating range from the small job shop through large industrial complexes which could support their own cogeneration facility. The large industrial complex should not be overlooked as a customer for district heating. There are many cases where the owner does not want to, or cannot, build his own heating plant, but has a substantial requirement for building heating and cooling, or process steam. These complexes can serve as an excellent line load around which the district heating system can be built.

The preceding discussion considered classes of buildings independently, assuming that the whole community was composed solely of either residential, commercial, or industrial types. This is not the case in the real world. A mix of classes occurs, which can allow the district heating system to serve areas of the community that could not be economically served when considered independently. For example, if the distribution system must be run through a single family residential area to a central business district or industrial customer, the incremental cost of adding the residential customers to the system is relatively small. Thus, the range of application is virtually unlimited, but is highly dependent upon the number and mix of customers in the service area.

VI. FUEL CONSIDERATIONS

It is not reasonable to expect to build a coal-fired plant next to the largest department store in the central business district. The logistic, aesthetic, and political problems would be insurmountable. On the other hand, it would not be prudent to design the district heating system around gas-fired cogeneration units and expect them to compete economically with existing gas-fired boilers and forced-air systems. The cogeneration units are the heart of the district heating system. They are the base-load plants. The choice of fuel must be economically, environmentally, and politically acceptable, since it must serve the heating needs of the community and provide electricity at a cost comparable with other base-load plants on the electrical system. This requires innovation on the part of the designer, both in the choice of fuel and combustion technology. For example, a heat-recovery boiler on a combustion turbine causes almost no reduction in the electrical output of the turbine, and therefore, can produce hot water or steam at a much lower cost than a boiler fired with the same fuel. However, from an electrical system point of view, these are peaking units because of their high cost of producing electricity relative to the large base-load plants.

The total cost to the user for district heating must be compared with the cost of conventional heating systems to determine if district heating can be offered at a competitive cost. Many studies have been performed which assume a single type of cogeneration plant, i.e., coal-fired boiler with back-pressure turbine, and a single type of conventional heating system, i.e., gas-fired forced air. These studies also assume that coal gasification or coal liquids are used to supply the conventional system. Unfortunately, the real world has a wide variety of in-building heating systems, and the district heating system must be economically competitive with existing fuels and may not be justified only by comparing it with a future possibility.

For in-building systems, the most economic alternative for the area can be used as a base. In most cases, this will be a natural gas system. If district heating can be shown to be competitive within this system, then it can compete with all other alternatives.

The district heating system presents a more complex problem. Piping costs will vary, based on local construction costs and difficulty of installation. Since a large fraction of the piping system cost is contained in the cost of cut and cover, which is nearly independent of the size of the pipe, the load density does not greatly affect the capital cost of the system, but it does greatly affect the cost to the customer. The mix of plants serving the district heating system will be determined by the number and type available for retrofit and the annual load dynamics for the system. The optimum economic advantage for retrofitting existing plants can be obtained by balancing the electrical and heat output. This is not always possible because of the size of the plant being retrofitted or because of differences in the heat and electrical load dynamics.

In the space available to this chapter, it is only possible to give cursory consideration to the economic analysis of district heating alternatives. A list of the key economic factors is shown in Table 3. The items for the district heating piping system are shown in the greatest detail because they are assumed to be the least familiar items to the reader, and because there are fewer possible variations in this system than in the power plant. The factors covered are for both steam and hot water systems. Obviously, the type of system being analyzed will determine the factors used.

A few points should be made about some of the items in Table 3. When an existing electric power plant is retrofitted for district heating, some electrical capacity is lost. Therefore, a replacement capacity charge ($/kW) is applied to compensate the electrical system for the lost capacity. Likewise, a replacement energy charge ($/kWh) is assessed against lost generation. Both of these factors are functions of the existing

TABLE 3

Economic Factors

Capital
Piping
- Pipe, valves, expansion, joints, manholes, steam traps, drains, etc.
- Pipe fitting materials
- Thermal insulation
- Radiography, electrical accessories, pipe flushing, etc.
- Civil work — trenching, refilling, and materials, routing, depth, type of construction, sizing
- Changes in other services
- Reinstatement
- Pumping stations
Plant — retrofit
- Equipment and installation
- Replacement capacity charge
- Replacement energy charge
Plant — new
- Equipment and installation
- Economy of scale
General
- Rate of return
- System life
- Effect on expansion plans
- User commitment/requirements
- System losses
- Taxes
- Government involvement
Operating expenses
- Fuel
- Maintenance — plant and system
- Operating
User system — conventional
- Equipment cost
- Age of equipment
- Fuel cost
- Maintenance cost
- Operating cost
District heating
- Heat exchanger
    Heating
    Hot tap water
    Cooling
- Radiators and distribution system
- Supply taps
- Pumps
- Maintenance cost
- Operating cost

electrical system. They do not apply to new cogeneration-plant construction since they are implicitly included in the design.

The commitment of users to the district heating system is probably the most important economic parameter. Since the cost of the piping network is the largest cost item for the system, total participation by all potential users in the service area is necessary to realize the optimum economic advantage. Part of this is the user requirements for district heating. If the user has high specific requirements, then larger piping is run, but there is no increase in the length of the piping network. Since the cost of increasing pipe size is very small compared to the cost of installing the network, additional savings can be realized on a $/MBtu basis.

The analysis to be performed is a comparison of a low capital/high operating-cost system (conventional gas and oil systems) with a high capital-cost/low operating-cost system (district heating). The user sees two separate cost items when comparing conventional and district heating systems: (1) the capital cost of the in-building system and (2) the operating, or "fuel", cost. The capital cost for the conventional system is the furnace or boiler, and the air conditioning and hot-tap-water heater. Duct work or piping in the building can be ignored for comparison with district heating because it is used in either system. Operating costs include "fuel" and maintenance costs. The "fuel" cost includes the capital cost of the extensive existing transmission and distribution network owned by the oil and gas companies. Even though the capital portion of a user's fuel bill can be as high as 80%, he sees this as the cost of fuel only. Maintenance is relatively low for these in-building systems, but should not be ignored in making the analysis since they are larger than for the district heating alternative. Depending on local ordinances, large in-building systems employing boilers may require a stationary engineer on duty to operate the system at an additional expense to the owner.

For the district heating system, the user's capital cost consists of the heat exchangers used in the building and the tie-in charge for hot water systems, or the tie-in charge alone for steam systems.

The cost of the heat exchangers is approximately equivalent to the cost for the conventional system in small buildings. In large buildings, the heat exchangers cost substantially less than the conventional system, and no stationary engineer is required for operation. The operating costs for district heating, again, are primarily "fuel" and maintenance in the eye of the user. The "fuel" cost is heavily weighted by the cost of the transmission and distribution system, but reduced by the low cost of energy from coal, refuse, or nuclear fuel. Thus, any reduction in length of piping per unit load or increase in load per unit length of piping has great impact upon the economic viability of district heating.

REFERENCES

1. **Collins, J. F., Jr.**, The history of district heating, *Dist. Heat.*, 62, 18, 1976.
2. **Anon.**, 1976 Statistics, Int. District Heating Assoc., unpublished data, 1976.
3. **Harboe, H.**, There is a case for energy conservation through district heating, *Energy*, Fall Quarter, 24, 1975.
4. **Oliker, I.**, Heat Exchanger and Thermal Storage Problems in Power Stations Serving District Heating Networks, paper presented at the American Institute of Chemical Engineers — American Society of Mechanical Engineers Heat Transfer Conference, Salt Lake City, August 15 to 17, 1977.
5. **Tourin, R. H.**, Combined Heat and Electric Power Generation for U.S. Public Utilities: Problems and Possibilities, presented at the Dual Energy User Systems Workshop, Electric Power Research Institute, Yarmouth, Maine, September 19 to 23, 1977.

Chapter 5

"TOTAL ENERGY" AND OTHER COMMUNITY SYSTEM CONCEPTS

W. R. Mixon

TABLE OF CONTENTS

I. INTRODUCTION

Cogeneration is a concept that can be applied in many forms. Previous chapters described the use of utility or industrial plants which provide thermal energy for process heat or district heating systems. Total Energy and other community system concepts generally utilize smaller generating plants that are located within a community to provide electricity, space heating and cooling, and potable-water heating. These systems can be (1) designed to meet the specific utility needs of the community, (2) installed in phase with community development, and (3) located near the users for economical distribution of thermal energy with hot- and chilled-water piping systems. For this application, a "community" could be a single building or any combination of buildings in the residential, commercial, and institutional sectors.

As the community size and system capacity increases, the distinction fades between Total Energy and the industrial or utility applications of cogeneration that include district heating. In fact, there are several terms used for community system concepts that are considered general and flexible enough to be synonymous with cogeneration. However, for the purposes of this book, we will focus on technologies appropriate for systems with generating capacities less than about 10 to 20 MWe. The corresponding community size would be characterized by utility demands equivalent to about 5000 to 10,000 dwelling units.

II. DESCRIPTION

The amount of primary fuel energy that is released to the environment at central station generating plants, and the potential conservation of energy which would result from the recovery and use of that energy, is widely recognized. Beneficial uses of waste heat, as low temperature condenser water, that are under investigation[1] include aquaculture and greenhouse heating. These applications are important for reasons other than energy conservation because fuel would not conventionally be consumed to supply heat for such purposes — at least not on the large scale proposed. Higher temperature applications, such as for district heating, require the use of cogeneration where, in steam-electric plants, electric generation is reduced in order to provide thermal energy at higher temperatures. Unfortunately, through the near future, the cost of thermal energy transmission may limit the application of central station cogeneration to installations that are relatively close to large population centers or appropriate industries.

Total Energy and other community system concepts offer another approach, the use of relatively small, dispersed cogeneration plants that are located near energy consumers. A Total Energy (TE) plant typically consists of several carefully combined components. There are a wide variety of commercially available options, and advanced technologies are under development that could be attractive for use in TE.[2] Using commercially available technology, combustion turbines or reciprocating engines would most likely be used to drive electric generators in the size range from 150 to about 5,000 kW.[3] Heat would be recovered from the exhaust of a gas turbine or from the exhaust and water jacket of a reciprocating engine in the form of low pressure steam or hot water.

The recovered heat could be used with a local district heating system to provide space and domestic-water heating. During the cooling season, recovered heat not used for water heating could be used with absorption-type refrigeration units to provide space cooling with a separate chilled-water distribution system. The absorption chillers would be sized and operated to take maximum advantage of available waste heat, and electrically driven compressive chillers would be used to provide the remaining cooling load. Auxiliary boilers provide required heat in excess of that available from waste heat.

The option of providing community energy services with a local TE plant offers possibilities for integrating other services, such as solid waste incineration with heat recovery and liquid waste treatment. Integration of liquid waste treatment could include the use of excess waste heat to facilitate sludge dewatering, sludge incineration, and the use of treated liquid waste for cooling towers or irrigation.

Heat rejection from the system may be by cooling tower or pond. The pond, in addition to serving as the heat rejection system, might be used as a source of water for fire protection, hold up of storm water, and in some areas of the country, it could also serve as a solar energy collector and water supply for a water-source heat pump serving low-density parts of the community.

With respect to the cogeneration concept, there is a basic difference between the steam-electric cycle, which is more appropriate for larger plants, and the technologies typically used for TE. The temperature of heat rejected from gas turbines, internal combustion piston engines, and many other prime-mover options for small systems is high enouh for practical use, and the temperature of heat recovery does not affect electric generating efficiency. Thus, recoverable thermal energy is actually waste heat that would otherwise be released to the environment.

Total Energy systems typically recover waste heat as steam or hot water at about 120°C, which satisfies several design criteria. With internal combustion engines designed for TE, the maximum recommended water jacket temperature is about 120°C, and exhaust gases should not be cooled below about 163°C in order to prevent condensation of corrosive compounds.[4] Single-stage absorption chillers are nominally rated for 12 psig inlet steam ($\sim$ 118°C),[5,6] and a hot water supply temperature as low as 93°C provides a reasonable temperature level and temperature difference for local district heating. Waste heat can be recovered at higher temperatures from gas turbine or engine exhaust gases, but the overall use of fuel energy would decrease as the waste-heat recovery temperature increases.

III. TERMINOLOGY AND OPERATING MODES

The concept of community systems can be applied with considerable flexibility in order to best meet the needs of particular communities. There are a number of terms commonly used to denote special types or operating modes of cogeneration.

The term *dual energy use systems* (DEUS), which is preferred by the Electric Power Research Institute,[7] appears to be synonymous with cogeneration, with the possible exception that DEUS specifically includes use of power-plant reject heat.

Total Energy usually refers to a cogeneration plant that is totally isolated from the local electric utility during normal operaton. The plant supplies all electric demands of the community and operates to follow the electric load. Heat recovered as a by-product provides part of the community's heating and cooling demands. TE plants may have emergency power connections with the utility, but typically include multiple components and excess capacity as necessary to ensure reliable service.

Another approach, termed Selective Energy (SE), is to operate the cogeneration plant with some combination of fossil fuel boilers to follow the community heating load. In this case, electricity may be considered a by-product. Generators are usually sized to meet only part of the electric load and the remainder is purchased from the local utility.

The possibility of active utility company participation with a utility intertie makes various other operating modes feasible. A SE system can then be designed for optimum overall efficiency; excess electricity would be absorbed by the utility, and the utility

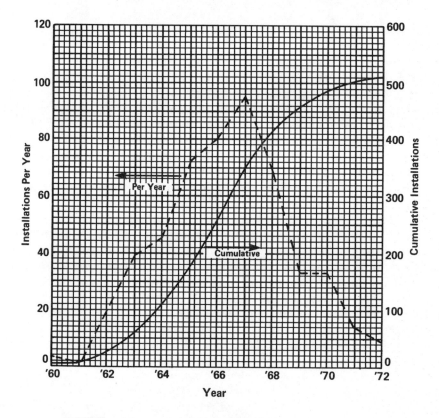

FIGURE 1.   History of reported Total Energy installations in the U.S.

could provide up to full backup. System operation could be quite flexible and arranged for the mutual benefit of all parties. The cogeneration system could even operate as a utility peaking plant and purchase utility electricity during off-peak periods. It is likely that the mode of operation of such a cogeneration system would vary over the life of the plant as energy economics and load profiles change.

IV. HISTORICAL DEVELOPMENT AND FEDERAL PROGRAMS

Onsite power generation and cogeneration was used in industry and large hospitals and commercial establishments around the turn of the century. These were primarily coal-fired steam-electric systems and most were abandoned as the electric utility industry developed. There are, of course, some notable exceptions. The Louisiana Station cogeneration plant built by Gulf States Utilities Company in 1930 is still operating successfully to furnish steam to adjacent industries, to generate part of the utility-supplied electricity, and to use some industrial waste products as a fuel supplement.

The modern history of Total Energy began in the early 1960s. The period from 1961 through 1967 highlighted a steady growth in the number of installations, the organization of the Group to Advance Total Energy (GATE) by gas utility companies, and Jeff Hunnicutt's publication of *Total/Energy*, which included technical and case-history material. Figure 1 shows the annual and cumulative number of installations reported by *Total/Energy*.[8,9] By the end of 1972, there was about 1,300 MW of generating capacity in 580 Total Energy plants. These installations typically serve new construction in which the owner of the plant was the primary or only energy customer. The growth of TE can largely be attributed to gas utility promotion as a means to increase and to improve the seasonal distribution of sales, but no owner would have

TABLE 1[10]

Sectorial Analysis of TE Survey

	Residential	Commercial	Industrial related	Total
In operation	15	59	114	188
Failed	7	38	65	110
Undetermined	11	25	99	135
Total	33	122	278	433

been interested if the plant had not been considered a profitable investment. The declining installation rate after 1967, which fell to less than 10 plants in 1972, was due to a number of factors, including changing energy economics and regulations, indications of trends to gas and oil shortages, and some unsuccessful examples of TE.

The lessons to be learned from the nation's experience with Total Energy are as yet inconclusive. Opponents of the concept point to the rapid decline and cessation of TE construction in the early 1970s and to cases of abandoned plants. Proponents can describe many TE plants that are still operating reliably and economically. The last TE plant directory[9] was published in January 1973 and, since that time, there has been no published record of new installations or of plants shut down. Research activities directed towards establishment of a credible data base from operating experience include:

- A survey, sponsored by the Department of Housing and Urban Development (HUD), of known TE facilities[10]
- Case studies of cogeneration facilities by Argonne National Laboratory for the Department of Energy (DOE)[11]
- A HUD funded demonstration of a residential TE system[12]

A. Survey Results

Results to date of the HUD-sponsored survey of 433 cogeneration plants listed in the 1973 *Total/Energy Directory and Data Book*[9] indicate that approximately 188 plants (43%) are in operation, 110 (25%) are no longer operating, and the status of the remaining 135 had not been determined. Table 1 shows the distribution of plants by user sector. Industrial and related applications account for the majority of plants, 64% of the total listed and 61% of those found operating. Residential/commercial applications accounted for 36% of the total listed and 39% of systems operating.

B. HUD Total Energy Research

One of the goals of HUD is to assist sound development of the Nation's communities. HUD's goals include "decent housing and a suitable living environment for every American family" at a cost they can afford. Space conditioning, power, and domestic hot water are included as elements of a decent home. This purpose was in mind when HUD began to examine the Total Energy concept and its application to residential housing in 1970 when the Operation Breakthrough sites were in the planning process. One initial conclusion was that there was insufficient operational information available to determine how effectively the TE concept could be applied in the residential sector. Therefore, the Office of Policy Development and Research of HUD funded the construction and instrumentation of a Total Energy Plant at an Operation Breakthrough site in Jersey City, N.J.

The site itself consists of 486 dwelling units in six mid- and high-rise buildings,

50,000 ft² of commercial space, a school and a swimming pool, all on a 6 ½ acre site. Electric power, hot water, and chilled water are provided from the TE plant which includes five 600 kW diesel engine generators, recovery of water jacket and exhaust heat in a primary hot water loop maintained from 82 to 110°C, two 4 MW auxiliary boilers, and two 546 ton absorption chillers. The engines and boilers burn # 2 fuel oil which is stored in three 25,000 gal underground tanks. The TE plant is completely automatic and normally operates unattended overnight and on weekends.

The plant was put into operation in 1974, and an analysis of 1 year of operation by the National Bureau of Standards (NBS) determined that a conventional central heating and cooling plant, purchasing power from the local electric utility, would have consumed 17.3% more resource energy.[12] Approximately 40% of the heat required at the site was the by-product of power generation which would otherwise have been wasted, and 80 to 90% of the recoverable waste heat was used. The cost of producing and delivering each service was approximately equal to that from conventional systems. The electric service reliability of the plant was reported to be 99.8%.

Another valuable benefit of the instrumented demonstration was the identification of several plant modifications that would reduce fuel consumption. Energy conserving measures included increasing absorption chiller efficiency by use of improved performance-monitoring techniques, installation of louvers on dry coolers to prevent natural convection heat losses, bypassing idle boilers to reduce heat loss to the equipment room, and elimination of boiler firing to provide equipment-room air conditioning. If all identified modifications were completed, the TE plant would consume about 25% less fuel than a conventional central heating and cooling plant.

C. Modular Integrated Utility Systems (MIUS)

The MIUS Program is a multiagency effort under the overall direction of HUD for the development, demonstration, evaluation, and commercialization of a new option for supplying community utility services.

The "energy crises" is a high priority national problem. However, developing communities face increasingly serious problems related to the adequate treatment and disposal of liquid and solid wastes, which must not be neglected. Many communities, faced with the task of upgrading existing waste disposal facilities and the growing need for new environmentally acceptable capacity, have resorted to building moratoriums to limit growth. The MIUS concept addresses these problems by integrating the solid- and liquid-waste treatment systems with a Total Energy system. For example, heat from the incinerator can replace the use of a boiler, or treated wastewater can be used for cooling tower makeup water for the power generation subsystem.

The primary goals of MIUS, using this total system approach to complete utility services, are to meet one or more of the following objectives:

- Conserve natural resources
- Reduce energy consumption
- Minimize environmental impact
- Provide utilities in phase with the demands of community development
- Eliminate the impact of local restrictions on waste treatment which delays housing construction
- Reduce total cost to the nation for utility services

The initial phase of the program included a survey and evaluation of major components and subsystems that could be used in MIUS concepts providing all utility services; comparative analyses of performance, economics, and other impacts of MIUS and conventional system operation; and development of analytic tools, such as the

ESOP computer program.[14] Technology evaluation studies completed by the Oak Ridge National Laboratory (ORNL), the National Aeronautic and Space Administration (NASA), and the National Bureau of Standards (NBS) indicated that MIUS could be assembled from commercially available components, and that several advanced technologies under development could be attractive for use in MIUS.[2] This phase of the program is documented in more than 30 technical reports, including a Generic MIUS Environmental Assessment,[15] a MIUS Technology Assessment,[16] and a compilation of abstracts[17] of all MIUS publications.

The second, or demonstration, phase of the program was to have a private sector design of a MIUS for a residential/commercial application. The objective of HUD was to determine the validity of concerns about institutional barriers, and to document the solution of such problems, thus providing a road map for future MIUS owners.

As a result of a competition, an award was made to the Interstate Land Development (ILD) Corporation for the design of a MIUS at St. Charles, Md. HUD allowed the developer a free hand in the MIUS design, only requiring that the developer be cost effective in satisfying a site-specific performance specification. The design process continued through September 1977, and the design and institutional reports were developed by ILD for release in 1978.

The demonstration satisfied several of the institutional goals of HUD. ILD, with the local utility, arrived at an arrangement by which both organizations could profit from a MIUS interconnected to the grid. In addition, the Maryland Public Service Commission ruled this MIUS to be exempt from power plant siting laws and state regulation.

The third phase of the MIUS Program is directed at dissemination of the MIUS information developed in Phase I and Phase II. Specifically, the activity planned will use workshops and seminars to provide MIUS information to key decision makers.

D. Integrated Community Energy Systems (ICES)

The Community Systems Program of the Office of Buildings and Community Systems, Conservation and Solar Energy, Department of Energy (DOE), is concerned with conserving energy and scarce fuels through new methods of satisfying the energy needs of American communities. These programs are designed to develop innovative ways of combining current, emerging, and advanced technologies into Integrated Community Energy Systems (ICES) that could furnish any, or all, of the energy-using services of a community. The key goals of the Community Systems Program then, are to identify, evaluate, develop, demonstrate, and deploy energy systems and community designs that will optimally meet the needs of various communities. A "4E" approach embodies these goals, namely:

- To conserve Energy
- To preserve the Environment
- To achieve Economy
- In the design and operation of human settlements (Ekistics)

The overall Community Systems effort is divided into three main areas; (1) Integrated Systems, (2) Community Planning and Design, and (3) Implementation Mechanisms. The *Integrated Systems* work is intended to develop the technology component and subsystem data base, system analysis methodology, and evaluations of various system conceptual designs which will help those interested in applying integrated systems to communities. Also included in this program is an active participation in demonstrations of ICES. The *Community Planning and Design* effort is designed to develop concepts, tools, and methodologies that relate urban form and energy utilization.

This may then be used to optimize the design and operation of community energy systems. *Implementation Mechanism* activities will provide data and develop strategies to accelerate the acceptance and implementation of community energy systems and energy-conserving community designs.

The Community Systems Program is quite active in all areas, with projects that include the following:

- Development of a consistent data base on the performance, costs, and other characteristics of the major components and subsystems applicable to ICES. Results are available as a series of Technology Evaluation reports.
- Development of automated procedures for ICES planning and design that include preparation of computer programs for system simulation and design optimization.
- Performing case studies of operating cogeneration systems serving communities which concentrate on the decision processes leading to implementation and the actual operating experience of that system.
- Implementing multiphase cogeneration demonstration programs that include a coal-using ICES at Georgetown University, four grid-connected ICES projects, and projects to demonstrate the application of district heating and cooling systems for communities through power plant retrofit.

V. ENERGY CONSERVATION POTENTIAL

Detailed comparative analyses of fuel energy consumption of Total Energy (the electric-thermal subsystem of MIUS) and conventional utilities were completed as a part of the MIUS Program.[16] It was assumed that the TE model was completely independent from conventional utilities and that each system provided equal "services" to a hypothetical, 720-unit, garden apartment complex having 60 two-story buildings clustered on 40 acres.

Major components of the TE model used for analyses are shown schematically in Figure 2. Multiple engine-generator sets with excess installed capacity were used to provide reliability, and the heat-recovery system supplied 116°C steam to single-stage absorption chillers and 93°C water to a district heating system. The absorption chillers were sized and operated to take maximum advantage of available waste heat, and electrically driven compressive chillers provided the remaining cooling load. Auxiliary boilers provided heat in excess of that available from waste-heat recovery. Both oil- and gas-fueled systems were analyzed.

Performance of the conventional electric utility model was based on a projected mix of large steam-electric plant additions for the year 1985 consisting of 31% coal-fired plants at 36.3% efficiency and 69% nuclear at 31.0% efficiency. Plants were assumed to be equipped with environmental control features and to operate base-loaded with a 70% capacity factor. The conventional building equipment models consisted of combinations of electric and gas- or oil-burning systems, with each model utilizing electricity from the conventional electric utility. The four models considered were

- Model C, a conventional district system with central boilers and compressive chillers
- Model D, air-to-air heat pumps in each apartment
- Model E, gas-fired boilers in each apartment building plus individual apartment air conditioners
- Model F, electric resistance heaters and air conditioner units in each apartment

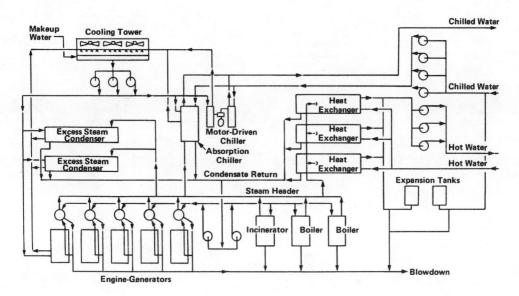

FIGURE 2.   Equipment-building schematic flowsheet for the MIUS thermal-electric subsystem.[16]

The results shown in Figure 3 are calculated fuel energy savings from MIUS use with respect to use of each of the conventional building service systems shown for five types of climate. For these analyses, the MIUS model did not include solid-waste incineration or thermal storage. Fuel energy consumption of conventional models included the on-site use of gas or oil and the fuel required to generate electricity. As expected, energy savings with MIUS were greater in cold climates and when compared to all-electric conventional systems. Significant savings were shown, however, for a year-around mild climate (San Diego) and for a climate with high cooling and low heating demands (Miami).

The energy savings resulting from this example are believed to be conservative. A detailed thermal balance of the MIUS model showed that one fifth or more of recoverable waste heat was not utilized, partly because of the noncoincidence of the time that waste heat was available and the time that heat was needed, and partly because of the simple garden-apartment model used. There are many possible, and perhaps more realistic, consumer models and techniques to optimize the total MIUS-consumer complex which would effect an increased utilization of waste heat. Success of an integrated total systems approach, such as the MIUS concept, requires careful planning and coordination and a concern for life-cycle costs. A definite advantage of these aspects is the provision of the mechanisms and incentives to realize the fullest potential for energy conservation.

Utilization of MIUS would effect a significant savings in the consumption of fuel energy, but the impact on a particular fuel resource depends on the type of fuel used in MIUS and in the alternative conventional systems used for comparisons. New conventional utilities were projected to use coal and nuclear fuel to generate electricity and some combination of gas, oil, or all-electric systems within buildings. On this basis, MIUS, using gas- or oil-fueled systems, would consume relatively more gas (or oil) than conventional utilities. Of course, if some mechanism should exist by which gas (or oil) to be used in MIUS could be reallocated from less efficient users of the same fuel, then the use of MIUS would result in a savings of that resource.

In order to avoid possible limitations in the use of MIUS due to shortages of oil and natural gas and to provide a more positive contribution to the conservation of

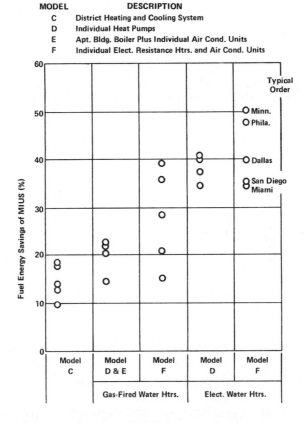

FIGURE 3.   Fuel energy Savings of MIUS over conventional
HVAC models serving 720 garden apartments. Based on a conventional central station generating efficiency of 33%.[16]

these premium fuels, HUD and the Office of Fossil Energy of DOE jointly initiated development of a coal-burning MIUS. This concept, under development at the Oak Ridge National Laboratory, utilizes a coal-fired, fluidized-bed combustion chamber coupled to a closed-cycle gas turbine to generate electricity.[18] As in current MIUS concepts, waste heat would be recovered and utilized for heating and cooling requirements.

REFERENCES

1. **Yarcsh, M. M., Ed.,** Proceedings of the National Conference on Waste Heat Utilization held in Gatlinburg, Tennessee, October 27 to 29, 1971, CONF 711031, published by Oak Ridge National Laboratory, Oak Ridge, Tenn., May 1972.
2. **Miller, A. J. and Samuels, G.,** *Technology Evaluation for MIUS,* paper no. 739149, *8th Intersociety Energy Conversion Engineering Conference Proceedings* held at the University of Pennsylvania, Philadelphia, Pennsylvania on August 13 to 17, 1973, published by American Institute of Aeronautics and Astronautics, New York, New York, 1973.
3. **Samuels, G. and Meador, J. T.,** Modular Integrated Utility Systems Technology Evaluation—Prime Movers, ORNL/HUD/MIUS-11, Oak Ridge National Laboratory, Oak Ridge, Tenn., April 1974.
4. **Segaser, C. L.,** Integrated Community Energy Systems Technology Evaluation of Internal Combustion Piston Engines, ANL/CES/TE-77-1, Argonne National Laboratory, Argonne, Ill., July 1977.

5. **Payne, H. R.,** Modular Integrated Utility System Technology Evaluation—Lithium Bromid—Water Absorption Refrigeration, ORNL/HUD/MIUS-7, Oak Ridge National Laboratory, Oak Ridge, Tenn., February 1974.
6. **Christian, J. E.,** Integrated Community Energy System Technology Evaluation of Central Cooling — Absorptive Chillers, ANL/CES/TE-77-8, Argonne National Laboratory, Argonne, Ill., August 1977.
7. Electric Power Research Institute, *Dual Energy Use Systems Workshop Summary held at Yarmouth, Maine, September, 1977,* Rep. No. EM-718-SR, Palo Alto, Calif., March 1978.
8. *Total/Energy, 1971 Directory and Data Book,* 8, (1), Total/Energy Publishing, San Antonio, Texas, 1971.
9. *Total/Energy, 1973-1974 Directory and Data Book,* 10(1), Total/Energy Publishing, San Antonio, Texas, January 1973.
10. **Orlando, J. A. and Friedman, N. R.,** Mathtech, Inc., *Total Energy Applications for Commercial and Residential Space Conditioning and Electricity Requirements,* in the Proceedings of the 5th Energy Technology Conference held February 27 to March 1, 1978, Government Institutes, Inc., Washington, D.C., April 1978.
11. **Faddis, R. J. and McClure, C. J. R.,** *The Performance of Operating Total Energy Plants,* 1st Int. Conf. on Energy Use Management Proc., Pergamon Press, New York, 1977.
12. **Hebrank, J., Hurley, C. W., Ryan, J. D., Obright, W., and Rippey, J.,** Performance Analysis of the Jersey City Total Energy Site: Interim Report, NBSIR 77-1243, National Bureau of Standards, Washington, D.C., July 1977.
13. **Rothenberg, J. H.,** "A Review of the HUD Total Energy Experience and the MIUS Program", paper presented at the 144th Ann. Meet., Am. Assoc. Advancement Sci., Washington, D.C., February 13, 1978.
14. **Hamil, R. G.,** Program Documentation for Energy Systems Optimization Program II (ESOP-II), Vol. 1, Engineering Manual, JSC 12625, Lockheed Electronics (contractor), National Aeronautics and Space Administration, Washington, D.C., March 1977.
15. **Mixon, W. R. and Row, T. H.,** Applications of Modular Integrated Utility Systems Technology, Final Environmental Statement, HUD-PDR-EIS-75-1F, Department of Housing and Urban Development, Washington, D.C., October 1975.
16. **Mixon, W. R., Ahmed, S. B., Boegly, W. J., Brown, W. H., Christian, J. E., Compere, A. L., Gant, R. E., Griffith, W. L., Haynes, V. O., Kolb, J. O., Meador, J. T., Miller, A. J., Phillips, K. E., Samuels, G., Segaser, C. L., Sundstrom, E. D., and Wilson, J. V.,** Technology Assessment of Modular Integrated Utility Systems, Vol. 1, Summary Report, ORNL/HUD/MIUS-24, Oak Ridge National Laboratory, Oak Ridge, Tenn., December 1976.
17. **Ryan, J. D. and Reznek, B.,** Editors, Abstracted Reports and Articles of the HUD Modular Integrated Utility Systems (MIUS) Program, NBS Special Publication 489, National Bureau of Standards, Washington, D.C., August 1977.
18. **Fraas, A. P., Anderson, T. D., Boegly, W. J., DeVan, J. H., Holcomb, R. S., Inouye, H., Lackey, M. E., Reed, S. A., Rittenhouse, P. L., Samuels, G., Segaser, C. L., and Tudor, J. J.,** Use of Coal and Coal-Derived Fuels in Total Energy Systems for MIUS Applications, Vol. 1, Summary Report, ORNL/HUD/MIUS-27, Oak Ridge National Laboratory, Oak Ridge, Tenn., April 1976.

Cogeneration Technology

SECTION 2

COGENERATION TECHNOLOGY

PREFACE

Several technological alternatives are available for cogeneration systems. Each of these has applicability for certain uses. For example, the approximate ratio of electric power to by-product useful heat varies from 0.05 kWe/lb/hr of steam with a steam turbine, to 0.25 kWe/lb/hr with a gas turbine system, to 0.5 kWe/lb/hr for a diesel cogeneration complex. Some alternatives (fuel cell systems) are capable of various ratios as the demand varies.

Traditionally, the use of a topping steam turbine to achieve shaft work while reducing the steam pressure to a lower, yet useful, level has been commonly considered when cogeneration is suggested. Indeed, a great deal of experience is available with this system. In the chapter which follows (Chapter 6) the characteristics of steam turbine systems are discussed and the use of topping (and bottoming) steam turbines is reviewed with respect to performance characteristics and economic factors.

More recent, but still extensive, use of gas or liquid-fuel turbine units with heat-recovery operations has been experienced. Low fuel and capital costs in the past have encouraged the operation of such systems. Increased fuel charges have made it desirable to increase the efficiency of operation by optimization of the overall cogeneration cycle. In Chapter 7, the gas turbine systems are discussed from design, operation, and thermodynamic points of view. Performance characteristics are predicted as functions of varying operating parameters to enable the effects of these parameters to be studied. An important question is the availability of alternate fuels for gas turbines, and these are reviewed, particularly with respect to coal as a turbine feedstock. Finally, a design case is presented to demonstrate the economic evaluation of such a system.

A somewhat less common, but still important, cogeneration concept involves the heat recovery from the exhaust of internal combustion engines. As with the other cogeneration systems, the reciprocating-engine unit has characteristic heat to power ratios, and thus, is useful for certain applications. The performance ratios of diesel engines typifying this type of cogeneration technology are given in Chapter 8 along with a discussion of coal burning as a fuel for diesel engines. A conceptual plant design is included to demonstrate the layout, performance, and economic analysis of a diesel cogeneration plant.

More advanced cogeneration systems involve alternate energy sources such as fuel cells, Stirling engines, organic Rankine cycles, and modified diesel and gas turbine concepts. The fuel cell is probably the most readily available system, and Chapter 9 gives the thermodynamics, principles, and operating performance of fuel cells which produce electricity directly while rejecting useful heat to a cogeneration system. Similar, less detailed analyses are given for Stirling engines, organic Rankine cycles, coal-burning diesel units, externally and directly fired gas turbines.

Chapter 6

STEAM TURBINE SYSTEMS FOR COGENERATION

Martin C. Doherty

TABLE OF CONTENTS

I. INTRODUCTION

Steam turbine-generators are the most common type of equipment used for the co-generation of electricity and heat. Steam turbines have been applied to cogeneration for over 70 years and have developed into a mature, proven technology. Many different types and sizes of steam turbines have evolved for use in cogeneration systems. These range in size from a few hundred kilowatts to over 200 MW. Most cogeneration steam turbines are in the range of 5 to 40 MW and have been installed in industrial factories with continuous manufacturing processes: pulp and paper mills; petroleum refineries; chemical plants; steel, aluminum, and copper mills; and food processing plants. There are also a number of co-located steam turbine cogeneration systems. In these cases, there are generally two or more firms involved. Steam and power are supplied "over-the-fence" from the power plant owner to the user's plant. Other cogeneration steam turbines have been installed in electric utility plants and institutions to serve district heating systems.

II. EQUIPMENT AND SYSTEMS DESCRIPTION

A. Steam Turbine-Generator Equipment

There are many types of steam turbines which have been developed for cogeneration systems. All supply useful heat in the form of steam to meet an external demand. Figure 1 schematically illustrates three types of noncondensing steam turbines. These all exhaust directly into a steam header. The straight noncondensing (SNC) units supply steam only at the single pressure of the exhaust. The single and double automatic-extraction units (SAENC and DAENC) supply steam at two or three different pressure levels.

The term "automatic extraction" implies that the extraction steam flow is automatically controlled (governed) to maintain the pressure of the header independent of the extraction flow. Figure 2 is a cross-section view of an SAENC. Notice that there are two sets of steam-control valves. The inlet-valve gear regulates the amount of steam admitted to the high-pressure section of the turbine. The extraction-valve gear regulates the amount of steam which passes to the low pressure section. The turbine governor operates both sets of valve gear in concert to control the pressure of the extracted steam and one other variable, such as exhaust pressure or power output. Figure 3 is an axial view cross-section of a typical turbine inlet valve. Multiple poppets are individually operated through cam-action lifts. This minimizes throttling losses for high efficiency over a wide range of steam flow. Figure 4 is a cross-section of spool-type extraction-valve gear. It is a variable restriction which is mounted downstream of the extraction opening.

The second class of cogeneration steam turbines exhaust into a condenser. Single (SAEC), double (DAEC), and triple (TAEC) automatic-extraction condensing steam turbines are shown schematically in Figure 5. The steam which expands to the condensing section of the turbine typically is reduced in temperature to less than 130°F. At this level, the energy in the exhaust is generally waste heat and not useful for recovery. Figure 6 is a cross-sectional view of a double automatic-extraction steam turbine. Notice that there are three sets of valve gear.

Straight condensing steam turbines are used for the generation of power only. These do not fall within the normal definition of cogeneration equipment.

B. Steam Condensers

The steam condenser is a shell-and-tube heat exchanger which is generally hung be-

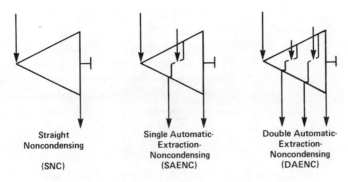

Straight Noncondensing (SNC)

Single Automatic-Extraction-Noncondensing (SAENC)

Double Automatic-Extraction-Noncondensing (DAENC)

FIGURE 1.   Types of noncondensing steam turbines.

FIGURE 2.   Cross-sectional view of a single automatic-extraction noncondensing steam turbine. (Courtesy of General Electric Medium Steam Turbine Department, Lynn, Mass.)

neath the condensing section of the steam turbine. The steam flows downward over the tubes, and cooling water is circulated through the tubes. The latent heat of the steam is rejected to the cooling water as it is converted from a vapor to a liquid. The condensed water is collected in the bottom in the hotwell. The steam condenser has three main functions:

- It collects the valuable high purity steam for re-use in the cycle.
- It is a pressure vessel which allows the steam to expand down to subatmospheric pressures. This extends the energy range of the steam which increases the amount of power generated per pound of steam flow.
- It is a heat exchanger which converts the exhaust flow from a vapor to a liquid. The liquid can be pumped back up to the inlet pressure at a very small fraction of the power which would be required to recompress the steam.

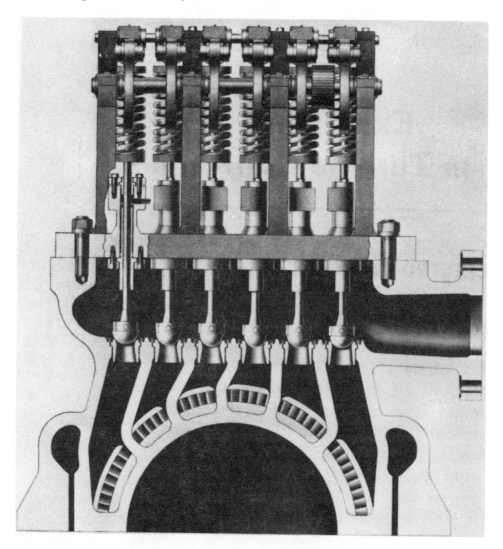

FIGURE 3.   Axial view of a turbine inlet-valve gear. (Courtesy of General Electric Medium Steam Turbine Department, Lynn, Mass.)

C. Cogeneration System Types

1. By-Product Power Generation (Topping Turbines)

Figure 7 shows two steam turbine cogeneration systems which generate by-product power. Figure 7A is a straight, noncondensing steam turbine supplied by a fossil boiler which provides steam at a single process pressure. In Figure 7B, a double automatic-extraction noncondensing steam turbine provides steam at three controlled pressures. In steam turbine systems, by-product power generation implies that power is generated by noncondensing turbine sections, either an extraction section or a noncondensing exhaust section. Both cases shown in Figure 7 would generally be operated in a pressure-governing mode. That is, the turbine inlet valves and the extraction valves would be regulated in order to maintain constant exhaust and extraction pressure. The amount of power generated would be a function of the process steam demands and, hence, is a by-product of supplying the process heat. In order to meet the total electric power demand, another source of power generation or a utility tie would be required.

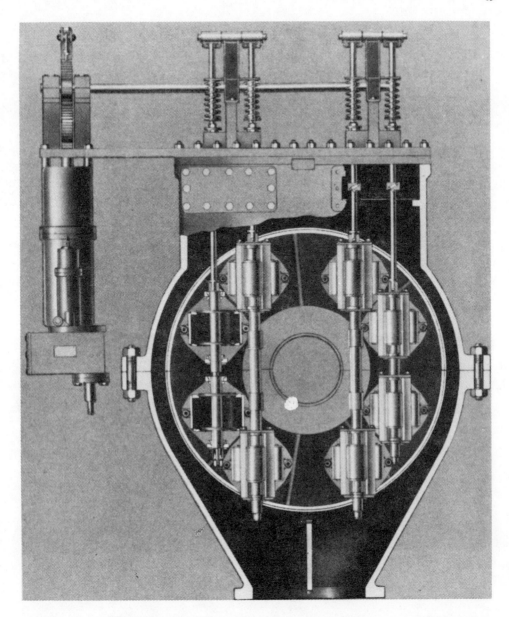

FIGURE 4. Cross-sectional view of a spool-type extraction-valve gear. (Courtesy of General Electric Medium Steam Turbine Department, Lynn, Mass.)

Most noncondensing steam turbines which operate on pressure governing are synchronized with the other source of power. The turbine-generator speed is locked into the system frequency, and therefore, speed is not a controlled variable during normal operation.

Steam turbine by-product power generation is sometimes called topping-turbine power generation. However, the term "topping turbine" has another definition as illustrated in Figure 8. When an old, medium pressure power plant is modernized by a new high pressure boiler, the older turbine-generators are frequently not replaced. A single new "topping turbine" can expand the steam from the new high throttle pressure to the old medium throttle pressure. Many industrial power plants have been modernized by this type of topping turbine.

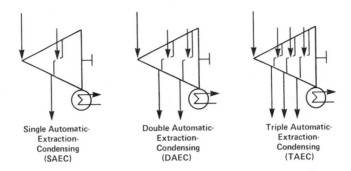

Single Automatic-
Extraction-
Condensing
(SAEC)

Double Automatic-
Extraction-
Condensing
(DAEC)

Triple Automatic-
Extraction-
Condensing
(TAEC)

FIGURE 5.   Types of automatic-extraction condensing steam turbines.

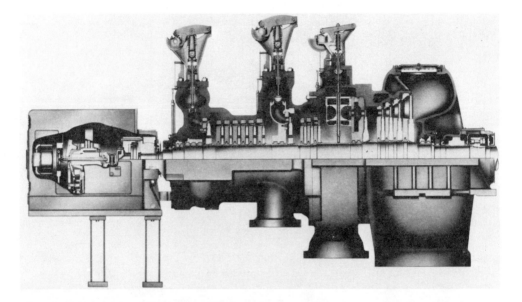

FIGURE 6.   Cross-sectional view of a double automatic-extraction condensing steam turbine. (Courtesy of General Electric Medium Steam Turbine Department, Lynn, Mass.)

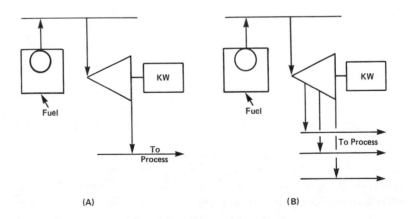

(A)   (B)

FIGURE 7.   Schematic diagram of steam turbine by-product power generation.

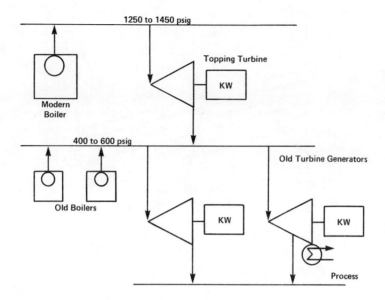

FIGURE 8. Schematic diagram of a topping steam turbine-generator.

A few other comments on the term by-product power generation are in order. Any heat engine, such as gas turbines or diesels, which supply useful exhaust heat are considered to generate by-product power. The power need not be electricity. By-product power can be generated by heat engines which drive pumps or compressors directly, even though it would be measured in terms of horsepower instead of kilowatts.

Finally, there can be exceptions to the relationship between by-product power and noncondensing steam-turbine power generation. Occasionally, when more electrical power is required than is available from existing sources in a particular industrial plant, additional steam can be expanded through a noncondensing steam-turbine section to meet the electricity demand. This excess steam is then either vented to the atmosphere or recovered in a separate heat exchanger which condenses the steam (a dump condenser). This steam does not provide any useful heat and is not considered to generate by-product power. This mode of operation is normally limited to transient conditions as it is a very inefficient use of energy.

2. Condensing Power

Automatic-extraction condensing steam turbines have the capability to supply process steam at a controlled pressure and to independently control the amount of electric power generated by the condensing section. Power generated by the steam which is extracted to the process is by-product power. However, the power generated by steam expanding to the condenser is not. It is simply referred to as condensing power. Of course, if the condensing steam turbine had no extraction, it would provide electric power only and no useful heat. It would then not qualify as a cogeneration system.

A great deal of energy burned in the boiler for condensing power is rejected to the surroundings through the condenser circulating water. Hence, the fuel is not used as effectively as in by-product power generation. Nevertheless, many cogeneration steam turbine systems do include some condensing power. Some of the reasons for including the condensing section are

- No external power is available from a utility.
- The energy in the steam is available at low cost, either from a by-product or waste fuel, or process-waste-heat recovery.

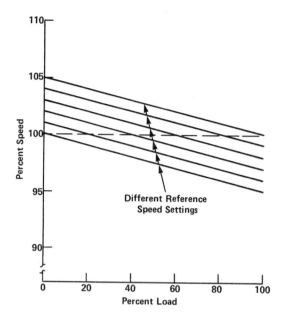

FIGURE 9.   Control action of a proportional or droop speed governor.

- Condensing power may be economic to shave peak electric demands and, therefore, reduce demand charges paid to the utility.
- When the industrial plant has several steam turbine-generators, normal operation would have a minimum amount of condensing power, but the condensing end can be used to full capacity as reserve during shutdown of another generator.
- When the industrial plant has several steam turbine-generators, a condensing section on one unit can simplify dispatching of power among the units.
- In joint utility/industrial cogeneration schemes, the condensing power is another source of generation which can be dispatched as the economics of the system demand.

There are several types of control modes available for the extraction-condensing steam turbine. In an isolated electric system, the governor would control extraction pressure and speed. Variations in extraction-steam demand or electric power demand would cause transient speed excursions. These excursions would be instantaneously corrected by regulating steam flow to the condenser.

When the extraction-condensing unit is synchronized on a large system, there would be no speed excursions (other than following the grid frequency which typically holds within 1/10 Hz). The most common type of control would be extraction-pressure control and speed/load control. The speed control mode of the governor would be a proportional or droop control loop. Its control action is illustrated in Figure 9. If the turbine is at 100% speed and zero load and an upset of 100% load is applied, the speed will droop by a value which is typically 5%. However, the speed in a synchronized unit is always at 100% and does not vary. Therefore, the amount of power generated will depend on the set point of the speed control loop. A knob on the governor can be adjusted to vary the reference speed setting and, therefore, the load carried by the unit. Hence, the term "speed/load" control. Of course, the speed/load set point can also be adjusted automatically by an external controller such as a tie-line control, or some master or supervisory control system.

Extraction-condensing units can therefore control extraction pressures while holding

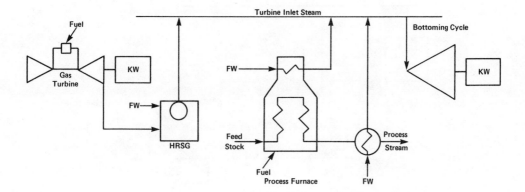

FIGURE 10.   Examples of heat sources for steam turbine bottoming cycles.

electric power output constant, or also vary the electric power in response to external signals.

Electrohydraulic or mechanical-hydraulic control systems have been developed to an advanced stage and have the capability of providing most of the automatic turbine-control functions required in the industrial power plant.

There is also an exception to the definition of condensing power. Some industrial processes and district heating systems have the requirement to heat large quantities of cold water. This water can be preheated by using it as cooling water in the turbine condenser. Therefore, the steam expanded to the condenser also provides useful heat and is considered by-product power. However, these types of applications are not typical as the quantity of water to be heated, in most cases, is a small fraction of the condenser cooling-water flow.

3. Bottoming Cycles

A bottoming cycle is generally considered one which utilizes waste heat from another source to generate power. Some waste heat sources are

- Gas turbine exhaust gas
- Fired process-heater exhaust gas
- Process streams which require cooling

These are illustrated in Figure 10. The 800 to 1000°F exhaust of a gas turbine is used to generate turbine inlet steam by a heat-recovery steam generator (HRSG). The process furnace stack can also be used to generate steam. In some processes such as ethylene and ammonia, process streams are heated to 1700°F in a furnace and are then cooled or quenched. This heat can also be used to generate turbine inlet steam.

It should be pointed out that by-product or topping-turbine cycles and bottoming cycles are not mutually exclusive. That is, a noncondensing steam turbine-generator can bottom one process heat source and at the same time top a lower temperature heat requirement. A bottoming cycle can also be a straight condensing steam turbine or an automatic-extraction condensing steam turbine, providing the heat source is a waste-heat stream such as shown in Figure 10.

Figure 11 illustrates another type of bottoming cycle. In this case, the process-heat recovery might be from a flue gas of only 500 to 600°F. This is too low to generate turbine inlet steam in this case, but adequate for generating 150 psig process steam. When the heat-recovery steam generation exceeds the process requirement, the excess steam can be admitted to the turbine-generator. Extraction-admission is another feature which can be specified for cogeneration steam turbines.

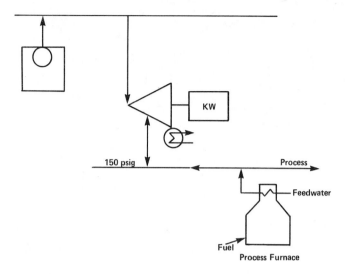

FIGURE 11.   Bottoming cycle by an automatic-extraction admission steam turbine.

Power generation cycles which rely on by-product fuels are similar to bottoming cycles. Some fuel sources are

- Forest products industries (pulp, paper, plywood, sawmills): wood waste, sawdust, bark, and black-liquor recovery boilers
- Petroleum and chemical industries: residual hydrocarbons, off gases
- Basic steel making: coke oven gas, blast furnace gas, BOP off gases
- Food processing: fibrous wastes such as bagasse in sugar refining
- Refuse disposal plants (trash)

One type of control mode available for steam turbines which are supplied solely by waste heat or by-product fuel boilers is initial pressure governing. The steam generated in the boiler might vary independently due to some other process. With an initial pressure governor, the turbine inlet valves would be controlled to hold the throttle pressure constant. The turbine would draw out of the header only that amount of steam that could be generated.

One final comment on the definitions of the various types of cogeneration systems. Most industrial plants which cogenerate encompass several of the system types. For example, bottoming cycles rarely exist in their purest form. They are generally supplemented by some fossil fuel steam generating capacity in order to independently control steam production. Many industrial powerhouses include automatic-extraction condensing steam turbine-generators which are supplied by waste heat or by-product fuel boilers and fossil fuel boilers. Therefore, a single unit often represents a bottoming cycle, by-product power generation or a topping cycle, condensing power cycle, and by-product fuel cycle.

III. ENERGY CONVERSION CONSIDERATIONS

A. Steam Turbine-Generator Efficiency

The accepted definition of efficiency of steam turbines recognizes that not all of the energy in the inlet steam is available for conversion to output power. The steam at the outlet pressure has energy which is not available even for a perfect (isentropic) turbine expansion. Therefore, the definition is expressed as:

TABLE 1

Sample Theoretical Steam Rate Table

Theoretical Steam Rates
in lb/kWh
Initial pressure = 850 psig.

Exhaust pressure	Initial temperature (°F)		
	800	825	850
Inches HgA			
2.0	6.685	6.580	6.478
4.0	7.174	7.058	6.944
Psig			
5	10.03	9.84	9.65
50	13.37	13.07	12.78
150	19.08	18.61	18.15
300	29.23	28.50	27.80
600	80.5	78.5	76.6

$$\text{Turbine Efficiency} = \frac{\text{Generator Output}}{\text{Available Energy}}$$

The available energy is the change in enthalpy from the inlet conditions to the outlet pressure for a perfect expansion process. The values of enthalpy can be read from steam tables or a Mollier diagram, or more conveniently, from a table of Theoretical Steam Rates (TSRs). The TSR is defined as:

$$\text{TSR (lb/kWh)} = \frac{3412 \text{ (Btu/kWh)}}{(\Delta H)_{isen} \text{ (Btu/lb)}}$$

A sample section of a TSR table is shown in Table 1. The TSR indicates the lb/hr of steam flow to produce 1 kW.

Cogeneration-type steam turbines generally have peak efficiency levels in the range of 70 to 80%. The Actual Steam Rate (ASR) is defined as the TSR divided by the turbine efficiency. Turbine output can readily be determined from:

$$\text{Turbine Output} = \frac{\text{Steam Flow (lb/hr)}}{\text{ASR (lb/kWh)}}$$

For example, assume 500,000 lb/hr is expanded through a turbine with 75% efficiency over an energy range from 850 psig, 825°F to 150 psig.

$$\text{ASR} = 18.61 \div 0.75 = 24.81 \text{ lb/kWh}$$

$$\text{Output (kW)} = \frac{500,000 \text{ (lb/hr)}}{24.81 \text{ (lb/kWh)}} = 20,150 \text{ kW}$$

Another important aspect of cogeneration steam turbine performance is the amount of heat in the extraction or exhaust. An exact calculation requires a rigorous analysis, but a shortcut method is available for simplified estimates. This shortcut is based on the approximation that the energy dissipated to the surroundings by generator losses, radiation, seals, and bearings is $2\frac{1}{2}\%$ of the generator output. An energy balance of the turbine-generator yields the following:

$$H_{out} = H_{in} - \frac{3500}{ASR}$$

where: H_{out} is the enthalpy at the exhaust or extraction and H_{in} is the inlet enthalpy.

In the previous example, the exhaust enthalpy could be approximated as:

$$H_{out} = 1410.6 - \frac{3500}{24.81} = 1269.5 \text{ Btu/lb}$$

The steam tables indicate that steam at 150 psig and 1269.5 Btu/lb enthalpy would have a temperature of 494°F.

In condensing steam-turbine cycles, the inefficiency of the internal expansion process (wheel efficiency) has a direct relationship on the amount of heat rejected to the surroundings in the condenser. However, in a noncondensing cogeneration steam-turbine section, the wheel efficiency simply determines the fraction of the energy converted to power and the amount remaining in the steam as useful heat. So the turbine efficiency in noncondensing steam turbines directly affects the amount of by-product power generated, but not the amount of losses to the surroundings. However, efficiency is important because Btu in the form of power are much more valuable than Btu in the form of process steam.

In many industrial plants, individual process units require some shaft power and some low-pressure steam. There is a tendency by some plant designers to deliver 600 to 850 psig steam to the unit and expand it down to lower pressures for process through relatively small (less than 2000 hp) mechanical drive turbines. Many of these are single inlet valve, single-stage units with turbine efficiencies of 40 to 50%. Although such drivers deliver by-product power, it is substantially less power than would be available if the steam was expanded through a large turbine-generator.

B. Boiler Efficiency

Boiler efficiency is defined as the heat added to the steam divided by the fuel consumption. Radiation and unaccounted for losses generally represent only $1\frac{1}{2}\%$ of input energy. Most of the losses occur in the exhaust stack. These losses can generally be reduced by trapping the heat in the exhaust gas in an air preheater or an economizer, or both. Typical boiler efficiencies are based on 300°F exhaust temperature. They are

* Natural gas: 84%
* Oil: 86 to 88%
* Coal: 84 to 88%

These efficiencies are based on the higher heating value (HHV) of the fuel. Lower heating values (LHV) are commonly used for gas turbine and reciprocating engine performance. The difference is that the LHV does not charge the conversion device for the energy required to vaporize the H_2O in the products of combustion.

C. System Efficiency

The system efficiency of steam turbine cycles which generate power only is generally called heat rate. The net heat rate of a power station is defined as:

FUEL UTILIZATION EFFECTIVENESS

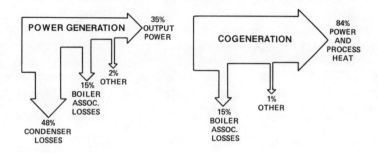

FIGURE 12.   Fuel utilization effectiveness of steam power plants. (From Wilson, W. B., ASME paper 78-WA/Ener-5, American Society of Mechanical Engineers, New York, 1975. With permission.)

$$\text{Net Heat Rate} = \frac{\text{Fuel Consumption}}{\text{Gross Output} - \text{Aux. Power}}$$

$$= \frac{\text{Fuel Consumption}}{\text{Net Output}}$$

Typical values of heat rate for large central steam stations are 9,000 to 10,000 Btu/ -kWh (HHV). This includes the inefficiency of the boiler, the turbine-generator, heat rejected to the condenser, and auxiliary power requirements.

In noncondensing steam-turbine cogeneration systems, the condenser losses are eliminated, and there are two useful outputs, heat and power. Typical fuel-utilization effectiveness of steam plants which generate power only and of cogeneration plants are illustrated in Figure 12. Only 35% of the energy is utilized in the separate power generation plant. By contrast, the cogeneration plant utilizes up to 84% of the input energy.

In order to more accurately define the efficiency of cogeneration systems, a concept called Fuel Chargeable to Power (FCP) has been developed. FCP can be calculated by first developing system performance data for a system which only produces heat. It should include the effect of boiler efficiency, blowdown, and auxiliary power requirements. This can be the actual performance of an existing plant, or estimated performance based on a new plant. Then, the performance of one or more cogeneration alternatives is developed.

FCP is calculated as follows:

$$\text{FCP} = \frac{\text{Fuel}_2 - \text{Fuel}_1}{\text{Gross Generation} - \text{PH Aux}_2 + \text{PH Aux}_1}$$

where: Fuel_2 is the cogeneration fuel consumption, Fuel_1 is the fuel consumption for the plant supplying heat only, PH Aux_2 is the auxiliary power required for the cogeneration alternative, and PH Aux_1 is the auxiliary power required for the plant supplying heat only.

If the amount of heat supplied to the process is reduced to zero, the definition of FCP is identical to the definition of heat rate.

FCP has wide acceptance as a measurement of cogeneration system efficiency. The FCP of steam turbine by-product or noncondensing power typically has a value of

4000 to 4200 Btu/kWh (HHV). The FCP of power produced by condensing sections of industrial-type steam turbines is in the range of 12,000 to 14,000 Btu/kWh. This is not as good as large central stations because the smaller industrial type cogeneration steam turbines are generally limited to 1450 psig throttle pressures without reheat, while the economics of large central station units permit throttle pressures up to 3500 psig with one or two stages of reheat.

When a cogeneration steam turbine includes extraction and a condensing exhaust, the FCP of the whole steam turbine can be approximated as a weighted average of the by-product power and the condensing power.

D. Optimizing the Cogeneration System

By-product power generation by steam turbines is one of the most efficient prime-mover energy-conversion technologies available. Fuel chargeable to power values of 4000 to 4200 Btu/kWh are equivalent to thermal efficiencies of 81 to 85%. Many plant designers and operators appreciate the benefit of by-product power generation, but fail to take full advantage of it. The optimum plant is not one which supplies all heat by cogeneration, but rather one which delivers the *maximum* amount of high efficiency power for a specific heat requirement. There are four parameters which should be considered in designing the optimum cogeneration plant:

- Inlet steam conditions
- Exhaust steam conditions
- Turbine efficiency (size)
- Feedwater heating cycle

Higher inlet steam conditions and lower exhaust conditions increase the energy range over which the steam expands. Each pound of steam delivered to process can deliver more power with an expanded energy range. Likewise, the turbine efficiency increases the amount of power per Btu supplied to process. Finally, regenerative feedwater heating is another technique of improving the amount of by-product power for a given heat load.

The effect of selecting the optimum design parameters for the cogeneration system can best be demonstrated by an example. Assume that 400,000 lb/hr of 50 psig steam is required in a process. Boiler efficiency is 88%, fuel is residual oil. The following five cases are to be evaluated:

Case	Throttle conditions (psig/°F)	Exhaust pressure (psig)	No. FW heaters	Turbine efficiency (%)
A	600/600	250	1	45
B	600/600	50	1	45
C	850/825	50	1	75
D	1250/900	50	2	75
E	1450/950	50	2	75

The low turbine efficiencies of Cases A and B are typical of small single-stage, single-valve mechanical drive turbines. The 250 psig back pressure of Case A might represent the plant designer not familiar with cogeneration who considers only the costs of piping for distributing steam to process. The smaller diameter 250 psig pipe is run throughout the plant and let down in pressure at the utilization equipment. Case C might represent typical industrial cogeneration practice prior to the 1974 fuel-cost escalation. Cases D and E, with higher throttle conditions and increased feedwater heating, should be considered in this plant size under current economic conditions.

TABLE 2

Performance and Annual Energy Costs of Example Cogeneration Plants

Case	A	B	C	D	E
Pressure (psig)	600	600	850	1250	1450
FCP (Btu/kWh)	4000	4000	4000	4000	4000
Net power (MW)	3.6	10.2	23.7	29.5	31.6
Fuel (M Btu/hr)	492	519	572	596	607
Purchased power (MW)	36.4	29.8	16.3	10.5	8.4
Annual energy costs ($ millions)					
Power	9.2	7.5	4.1	2.6	2.1
Fuel	10.3	10.9	12.0	12.5	12.7
Total	19.5	18.4	16.1	15.1	14.8
Savings	Base	1.1	3.4	4.4	4.7

Note: Basis: 400,000 lb/hr at 50 psig, 8400 hr/yr, 40 MW total power required, 3.0¢/kWh power cost, $2.50/M Btu fuel cost.

A summary of performance calculations and annual energy costs for all cases are shown in Table 2. Notice that the fuel chargeable to power for all cases is 4000 Btu/kWh. However, the net power generated varies from 3.6 MW in Case A to 31.6 MW for Case E. The effect of this large increase in high efficiency by-product power can best be appreciated by the annual energy cost calculations also shown in Table 2. These are based on a total power requirement of 40,000 kW, 8400 hr/yr of operation, $2.50/M Btu fuel cost and 3.0¢/kW power cost.

Notice the substantial annual savings which can be achieved with an optimum cogeneration system. Reducing the back pressure from 250 psig to 50 psig saves over $1 million/year. An additional $2.3 million are realized when a large, efficient turbine-generator is used instead of many small mechanical drive turbines. Finally, higher throttle conditions and increased feedwater heating result in total annual savings of almost $5 million as compared to Case A.

The final decision on equipment for a cogeneration system is not based only on fuel chargeable to power and annual energy costs. Capital costs and other operating costs must also be considered. The next section treats the subject of the economic evaluation of a cogeneration system.

IV. ECONOMIC CONSIDERATIONS

A. Introduction

Steam turbine-generators can produce substantial savings of energy in cogeneration systems. However, these plants require greater capital expenditures than plants which supply only low pressure process steam. This incremental capital investment must be justified by the reduced annual operating costs which result. Generally, a capital budgeting decision analysis is made to determine if the venture meets the corporation's requirement for return on investment or payout. Likewise, any additional investment for a larger or more efficient cogeneration alternative vs. a smaller system must also meet management's requirements for return on investment or payout.

The first step in making an economic evaluation of cogeneration systems is normally

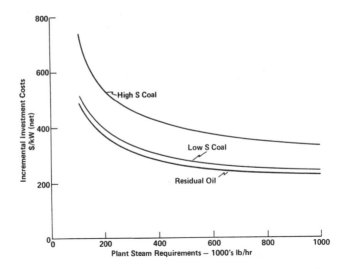

FIGURE 13.   Incremental investment costs of a cogeneration plant vs. plant size.

to determine the course of action which involves the minimum capital expenditure. This is the Base Case. For a new plant, this would be low pressure boilers for process heat and no cogeneration. For an existing plant, the Base Case might be the "do nothing" alternative.

Several cogeneration alternatives might also be proposed. These might include various throttle pressures, the use of a condensing section, or a combined steam and gas turbine. The Base Case and the cogeneration alternatives should then be developed to determine estimated performance data and equipment ratings. The next step would be to convert these data into terms of cold cash. That is, initial investment cost estimates for each alternative and annual operating costs. When these data have been developed, then a return or payout evaluation is made. If several cogeneration alternatives meet management's capital budgeting criteria, then each increment of additional investment should be evaluated. The final decision would be to select the alternative with the highest increment of investment which meets the return or payout criteria.

B. Investment Cost Data

In evaluating a cogeneration system, the incremental investment costs due to cogeneration should be the focus of the evaluation. Figure 13 illustrates typical values of incremental investment cost estimates for a range of plant sizes and fuel types. These data represent the differences in cost between a low pressure process-boiler plant vs. a cogeneration plant with steam conditions of an 850 psig, 825°F turbine inlet, and 50 psig exhaust. All cases include 50% spare boiler capacity. The cogeneration cases include one straight noncondensing steam turbine. The high-sulfur coal cases include the costs of a particulate removal system and a flue-gas desulfurization (FGD) system. The low-sulfur coal and residual oil cases include the costs of only the particulate removal system.

Notice that these costs exhibit a substantial economy of scale penalty for plant sizes less than 200,000 lb/hr. In the larger plant sizes, the incremental investment costs are $200 to $300/kW, with an additional $100/kW adder for FGD systems.

Investment costs in terms of $/kW will vary significantly with steam conditions. For example, a 200 psig process pressure instead of 50 psig would increase investment costs by approximately 50%.

TABLE 3

Premises of Economic Evaluation of Example Cogeneration Plants

Category	Premise
Annual operation	8400 hr/yr
Fuel costs	
Residual oil	$2.50/M Btu (HHV)
High-sulfur coal	$1.25/M Btu (HHV)
Purchased power cost	Sensitivity to a range of 2.5, 2.75, 3.0¢/kWh
Annual maintenance cost	2½ % of investment
Operating labor	
Process boiler cases	Base
Cogeneration cases	Base + $300,000/yr
Plant size	400,000 lb/hr process steam load

Note: Cost data from 1979.

TABLE 4

Summary of Annual Operating Costs and Economic Evaluaton of Example Cogeneration Plants

	Base case	850 psig	1250 psig	1450 psig
Fuel consumption (M Btu/hr)	476	571.6	595.8	606.8
Net power (kW)	(440)	23,720	29,500	31,600
Annual costs ($1000)				
Fuel	10,000	12,000	12,500	12,750
Power	100	(5500)	(6800)	(7300)
Labor	Base	300	300	300
Maintenance	300	460	530	550
Total	10,400	7260	6530	6300
Annual savings	Base	3140	3870	4100
Gross payout (years)	Base	2.07	2.40	2.41
DRR (%)	Base	29.3	25.8	25.9

C. Annual Operating Costs

The incremental investment costs for a cogeneration system are justified by the reduced annual operating costs which result. The focus of developing annual operating costs should be on any differential costs due to cogeneration. Normally these include costs of fuel, power, labor, maintenance, cooling water, and boiler makeup water.

To illustrate these calculations, evaluations of a 400,000 lb/hr plant size have been developed for residual oil and high-sulfur coal. In addition to the 850 psig cogeneration system developed for Figure 13, plants with throttle pressures of 1250 psig and 1450 psig have also been considered. The premises used in evaluating annual operating costs are shown in Table 3.

An example calculation of the annual operating costs for the residual fuel case with a purchased power rate of 2.75¢/kWh is shown in Table 4. Also included in this table is a summary of calculations of gross payout and discounted rate of return (DRR) on the incremental investment for the cogeneration cases. Gross payout is defined as:

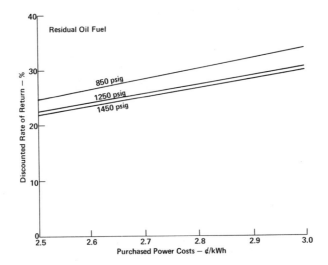

FIGURE 14.   Discounted rate of return on an oil-fired cogeneration plant vs. purchased power costs.

$$\text{Gross Payout} = \frac{\text{Incremental Investment}}{\text{Gross Annual Savings}}$$

The DRR calculations were made using the after-tax discounted cash flow analysis method. The parameters used in these calculations were

- Federal income tax rate — 48%
- Investment tax credit — 10%
- Depreciation — 28 years, sum-of-the-years method
- Investment life — 28 years
- Local property taxes and insurance — 2.5%

The typical range of minimum acceptable DRR values for industrial firms is in the range of 15 to 25%. All three cogeneration alternatives shown in Table 4 would pass such a criteria. The next step would be to determine DRR values for the incremental investment for the higher pressure cases vs. the 850 psi cogeneration case. In this example, the resulting DRR values would be 17 to 18%. For some corporations, the higher pressures would have an acceptable return, for others it would be unacceptable.

Figure 14 illustrates the DRR values for oil fuel for the range of purchased power costs. The same data for the high-sulfur coal case are shown in Figure 15. It is obvious that the return on a cogeneration system is sensitive to a number of factors.

Another important parameter not illustrated here would be the hr/year of operation. When a cogeneration system is operated at base load to serve continuous heat and power requirements, it generates maximum savings in annual operating costs. When the cogeneration system serves noncontinuous loads (5 or 10 shifts per week) and/or seasonal cyclic loads, the annual savings will be diminished. Therefore, cogeneration systems in these applications are more difficult to justify.

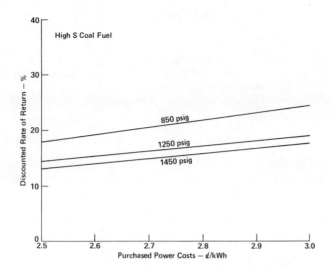

FIGURE 15.   Discounted rate of return on a coal-fired cogeneration plant vs. purchased power costs.

V. CURRENT TRENDS IN STEAM TURBINE COGENERATION SYSTEMS

For many decades, the majority of cogeneration systems have been located in industrial plants with unit sizes ranging from 5 to 40 MW. Throttle pressures have been typically in the range of 600 to 1450 psig.

The energy saving potential of cogeneration has been recognized by the U.S. government, and there is a trend to stimulate cogeneration by increased tax incentives and changes in regulatory requirements. It is anticipated that tax incentives will stimulate many industrial cogeneration applications with marginal return under current tax law.

The steam turbine cogeneration market is also expected to grow in unit ratings both in the smaller and larger ends of the traditional 5 to 40 MW range. In the smaller range, cogeneration is under consideration more and more in commercial and institutional applications. At the larger end of the spectrum, more electric utility participation is anticipated in plants which are co-located with one or more energy-intensive industrial plants. Such plants with outputs up to 450 MW have been in operation for over 20 years, but these have been the exception. Throttle pressures for these plant sizes are in the range of 1800 to 2400 psig.

Another trend is in the area of low-level-energy recovery. Many process waste-heat streams can be used to generate steam from 250 psig down as low as 5 psig. Even though these plant types incur high investment costs in terms of $/kW, annual operating costs are at a minimum due to the absence of fuel costs.

Finally, steam turbine-generator technology is being utilized for advanced cogeneration systems using other fluids such as isobutane or freon. These systems have the advantage of generating condensing power with reduced rejection of heat to the atmosphere.

REFERENCES

1. **Obert, E. F.,** *Thermodynamics,* McGraw Hill, New York, 1948.
2. **Salisbury, J. K.,** *Steam Turbines and Their Cycles,* John Wiley & Sons, New York, 1950.
3. **Spencer, R. C., Rossettie, P. G., and McClintock, R. B.,** *Theoretical Steam Rate Tables,* American Society of Mechanical Engineers, New York, 1969.
4. **Anon.,** Steam Turbine Generators for Industrial Applications, Report GEA-8510, General Electric Company, Lynn, Mass.
5. **Doherty, M. C.,** Selection of Boiler Steam Conditions for Industrial Power Plants, ASME Paper 75-IPWR-12, Am. Soc. Mechanical Engineers, Industrial Power Conference, Pittsburgh, Pa., May 19 to 20, 1975.
6. **Wilson, W. B. and Kovacik, J. M.,** Selection of Tubine Systems to Reduce Industrial Energy Costs, API Paper 18-76, Am. Petroleum Institute 41st Midyear Meet., Los Angeles, Calif., May 1976.

Chapter 7

COGENERATION SYSTEMS USING COMBUSTION (GAS) TURBINES

J. R. Hamm

TABLE OF CONTENTS

I. COMBUSTION TURBINE TECHNOLOGY

The substantial effort that has been expended on the development of the aviation gas turbine, starting during World War II and still going on, brought about the rather rapid development of stationary gas turbines, first for industrial and then for utility applications. Improved materials and the application of expander cooling techniques have resulted in an increase of expander inlet temperature from the initial 1200 to 1300°F level to current state-of-the-art levels in the 2000 to 2100°F range. Capacities of shop-assembled units with current temperatures are approaching the 100 MW level.[1] Operation in the 2500 to 3000°F temperature range is being projected.[2]

There are two primary Brayton-cycle categories, the open and the closed. A generalized model of the open-cycle gas turbine power system is shown in Figure 1. Specific configurations represented include the following:

1. Simple cycle
2. Recuperative simple cycle
3. Compound cycle with intercooler and reheat options
4. Simple cycles with waste heat boiler for steam injection
5. Combined Rankine and Brayton (simple or compound) cycles with fired and unfired heat-recovery steam generators

A generalized model of the closed-cycle gas turbines is shown in Figure 2. In the closed-cycle system, the working medium is at an elevated pressure which gives high volumetric power output. Heat is added to the working fluid through a metallic surface. It is possible, therefore, to use a working fluid other than air. Helium is the preferred working fluid for high performance.

The closed system permits the use of solid fuels such as coal in addition to gaseous and liquid fuels. Power output can be modulated by varying the pressure level. Thermodynamic performance is independent of cycle pressure level.

Other power systems that employ gas turbines are the Velox boiler[3] and the supercharged boiler[4] represented generically by Figure 3. The Velox boiler has no net output from the gas turbine, but the supercharged boiler does. In the latter case, the economizer would be reduced or eliminated.

A family of gas turbine cycles that may have a near-term potential for burning coal is that which uses indirect heating of the working fluids. The exhaust-heated cycle (shown in Figure 4) is a concept that was first investigated experimentally in the late 1950s at McGill University.[5] More recently, it has been considered as part of the MIUS program being carried out by Oak Ridge National Laboratory.[6] Figure 4 uses fluidized-bed combustion of solid fuel to fire the air heater. Provision is made for part of the exhaust air to by-pass the fluidized-bed air heater. Other types of combustors could be used with solid fuel or with dirty liquid fuels such as residual oil or coal-derived liquids.

An optional arrangement is that shown in Figure 5, where part of the gas-turbine working fluid is directly heated in a fluidized-bed combustor, and part is indirectly heated in tubes immersed in the fluidized bed. This concept is being investigated by Curtiss Wright under a DOE contract.[7]

The majority of the gas turbines produced by U.S. manufacturers in the last 30 years for industrial and utility applications have had simple open cycles without recuperators. Typically, the industrial units have been mechanical drives, e.g., natural gas compressor drives. Initially, utility applications were for peaking service. Recently, a significant number of combined-cycle systems have been placed in intermediate or base-

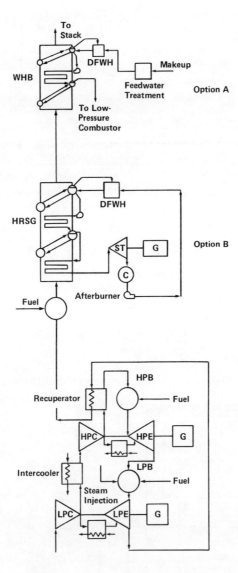

FIGURE 1.   Generalized model of open-cycle
gas-turbine power system.

load utility service.[8] No compound-cycle or closed-cycle plants have been installed in the U.S.

The use of the supercharged boiler in the U.S. has been limited to a few naval vessels where the size reduction of the boiler offered particular advantages.[9]

Most of the gas turbines operating in the U.S. burn either natural gas or No. 2 fuel oil. A few units have been operated on residual oil, and one unit has been operated successfully on blast-furnace gas.[10] Coal has been burned in two experimental gas turbines in the U.S. During the period from 1945 to 1965, the LDC turbine burned pulverized coal in an integrated combustor, first at Dunkirk, N.Y., and later at Morgantown, W.Va.[11] Economic feasibility was not demonstrated because of erosion problems. Combustion Power Company has operated an experimental unit on coal using an external fluidized-bed combustor.[12] This unit was originally operated on refuse-derived fuel.[13] Combustor-rig tests have been carried out on both actual and simulated low-Btu fuel gas[14] and on coal-derived liquid fuels.[15]

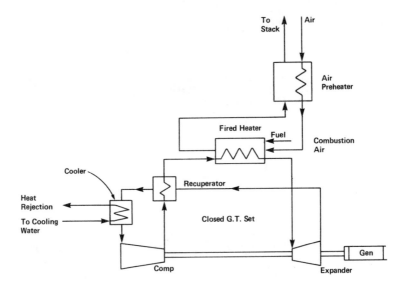

FIGURE 2.   Generalized closed-cycle gas-turbine system schematic.

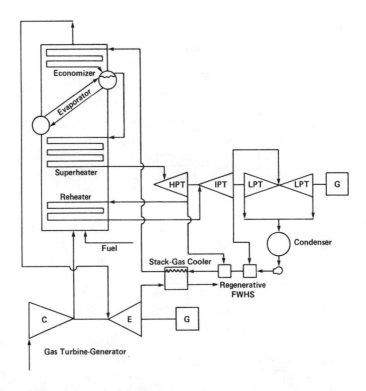

FIGURE 3.   Generalized model of Velox/supercharged boilers.

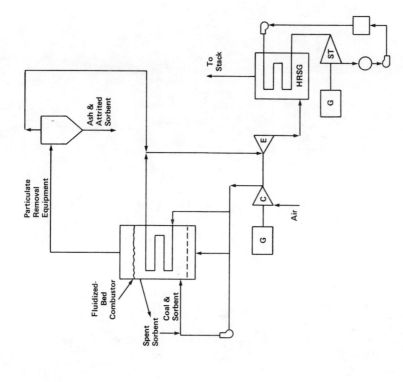

FIGURE 5. Combined-cycle system utilizing fluidized-bed combustion with indirect heating of part of the working fluid.

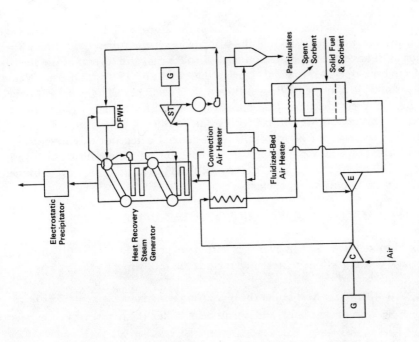

FIGURE 4. Solid-fuel-fired exhaust-heated combined-cycle power system.

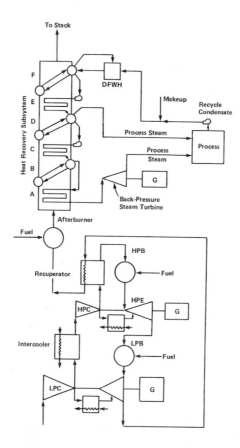

FIGURE 6.   Generalized model of open-cycle
gas-turbine cogeneration system.

Compound-cycle gas turbines are rather common in Europe.[16] A number of closed
cycles have also been installed.[17] While the principal fuel used there is distillate fuel
oil, numerous units have been operated on blast-furnace gas, and several of the closed-
cycle units have been operated on coal. During the 1930s and 1940s, numerous Velox
boilers were installed in Europe by Brown Boveri.[18] STEAG has a supercharged boiler
installed at Lünen, West Germany that operates on low-Btu fuel gas from fixed-bed
Lurgi gasifiers.[19] This unit, however, has never attained full operational status.

The Australian Department of Minerals and Energy carried out extensive experiments on gas turbines burning pulverized brown coal during the period from 1963 to
1972.[20] Here, again, commercial status for the coal-fired gas turbine was never attained
because of erosion and deposition problems.

II. CHARACTERIZATION OF COGENERATION SYSTEMS

A generalized model of the open-cycle gas turbine cogeneration system is shown in
Figure 6. Specific configurations represented include the following:

1. Simple cycle with waste-heat boiler (heat recovery subsystem elements DEF)
2. Simple cycle with recuperator and waste-heat boiler (heat recovery subsystem
    elements DEF)
3. Compound cycle with waste-heat boiler (heat recovery subsystem elements DEF)

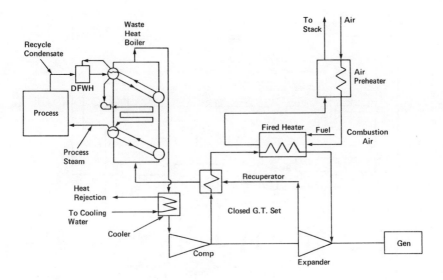

FIGURE 7.   Generalized closed-cycle gas-turbine cogeneration system model.

4.   Combined simple gas turbine cycle-steam cycle with heat recovery steam generator and back-pressure or extraction steam turbine (heat recovery subsystem elements ABCF)
5.   Combined simple gas turbine cycle-steam cycle with modified heat recovery
steam generator and back-pressure or extraction steam turbine (heat recovery
subsystem elements ABCDEF)

In all cases firing of the heat recovery steam generator or the waste-heat boiler is
optional, permitting the ratio of process heat to electrical energy to be varied.

A generalized model of the closed-cycle gas turbine cogeneration system is shown
in Figure 7. The design ratio of process heat to electric power can be changed by
varying the effectiveness of the recuperator. For a given design configuration, the
quantity of electrical power and process heat can be modified by varying the pressure
in the closed loop.

A generalized model of a supercharged/Velox boiler for cogeneration applications
is shown in Figure 8. The design gas-turbine power can be varied from zero upward
by changing the temperature of the combustion products leaving the supercharged
boiler. Waste-heat boilers can be designed to be either fired or unfired. With a fired
waste-heat boiler design, the ratio of process heat to electrical power can be varied by
modulating the fuel rate to the afterburner.

In the following pages, a number of cogeneration systems employing gas turbines
are characterized. This characterization is facilitated if the cogeneration system is divided into two subsystems, (1) the energy conversion subsystem and (2) the process-
heat utilization subsystem. An example is shown in Figure 9. This is so because each
of the several energy conversion subsystems and process-heat utilization subsystems
that are to be considered has its own unique characteristics which can be best demonstrated alone rather than in combination with those of the interfacing subsystem.

There are numerous possible configurations for process-heat utilization subsystems
and a great variety of operating variables for each configuration. Figure 10 shows a
generalized model of process-heat utilization subsystems that are applicable to back-
pressure or extraction steam turbines and waste-heat boilers.

Table 1 describes four specific configurations of this process-heat utilization model

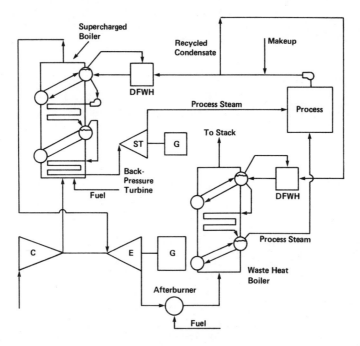

FIGURE 8.   Generalized model of a supercharged/Velox boiler for cogeneration application.

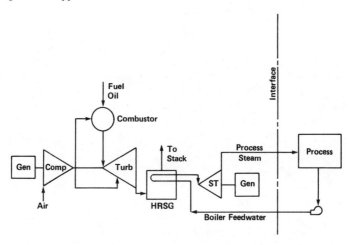

FIGURE 9.   Oil-fired combined-cycle cogeneration system.

that can be applied to cogeneration systems using combustion turbines. Configuration
A and B are for use with back-pressure or extraction steam turbines, while configurations C and D are for use with waste-heat boilers. In all cases, the boiler feedwater is
at process-steam supply pressure. Configurations A and C use the heater shown to
heat the boiler feedwater to the saturation temperature. In configurations B and D,
the temperature of the boiler feedwater is less than the saturation value.

Whenever process steam arrives at the point of use with a significant amount of
superheat, it will probably be desuperheated prior to use. Figure 10 includes a desuper-
heater that uses recycled condensate. The models with desuperheating are thermodynamically equivalent to those without.

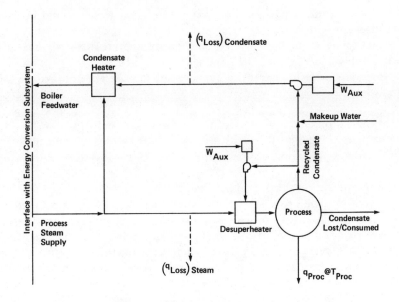

FIGURE 10. Generalized model of process-heat utilization subsystem.

TABLE 1

Specific Configurations of Process-Heat Utilization Subsystem Models

Configuration	Process-steam supply conditions	Boiler feedwater conditions
A	Based on applicable turbine expansion lines	p = Supply pressure t = t_{sat}
B	Based on applicable turbine expansion lines	p = Supply pressure t < t_{sat}
C	Dry and saturated at supply pressure	p = Supply pressure t = t_{sat}
D	Dry and saturated at supply pressure	p = Supply pressure t < t_{sat}

In the generalized process-heat utilization subsystem model, consideration is given to the following factors:

• Heat loss from the process-steam supply line
• Pressure drop in the process-steam supply line
• Loss of steam or condensate due to leakage or consumption in process
• Makeup of lost condensate
• Heat loss from the recycled condensate line
• Pressure drop in the recycled condensate line
• Pump work required to return the condensate to the process-steam supply pressure

Parametric calculations were made to characterize the process-heat utilization subsystem models in terms of the process-steam temperature at the point of use; thermal, pressure, and material losses; and the process-heat utilization factor. Assumptions made were as follows:

- The steam turbine expansion line is typical of that which would be used in an industrial application.
- Thermal losses from the recycled condensate line are 10% of those from the steam supply line.
- The pressure drops in the recycled condensate line are 10% of those in the steam supply line.

The heat utilization characteristics of models A and B and of C and D were determined to be thermodynamically identical. The manifestations of their differences are found in the characterization of the energy conversion subsystems with which they interface.

Process-heat utilization subsystem configurations A and B are characterized in Figures 11 through 15. The process-heat utilization factor is shown as a function of the process-heat temperature (212 to 400°F) and thermal losses (0, 50, 100, 150, and 200 Btu/lb) for two values of the process-steam pressure drop (0 and 0.3) and five values of the condensate recycle fraction (1.0, 0.75, 0.5, 0.25, and 0).

The ratio of subsystem auxiliary power to gross process heat is shown as a function of the process-heat temperature (212 to 400°F) for five values of the condensate recycle fraction (1.0, 0.75, 0.5, 0.25, and 0) and four values of steam supply-line pressure loss (0, 0.1, 0.2, and 0.3), is negligible over the range of parameters.

Process-heat utilization subsystem configurations C and D are characterized in Figures 16 through 20. The process-heat utilization factor is shown as a function of process-heat temperature and thermal losses for a range of steam-line pressure losses and the condensate recycle fraction. The ratio of subsystem auxiliary power to gross process heat for the waste heat-boiler models are shown in Figures 19 and 20. Here, again, the auxiliary power requirements for the process-heat utilization subsystem are negligible.

A number of energy-conversion subsystems using gas turbines have been characterized for cogeneration applications. These have been divided into two categories, state-of-the-art and advanced. The state-of-the-art systems considered are as follows:

- Oil-fired gas turbine with a waste-heat boiler (see Figure 21)
- Oil-fired combined cycle with a back-pressure steam turbine (see Figure 22)
- Coal-fired closed cycle (see Figure 23)
- Coal-fired exhaust-heated cycle (see Figure 24)

The advanced systems considered are as follows:

- Coal-fired gas turbine using fluidized-bed combustion with a waste heat boiler (see Figure 25)
- Coal-fired combined cycle using fluidized-bed combustion (see Figure 26)
- Combined cycle with integrated low-Btu gasification of coal (see Figure 27)

The characteristics of the oil-fired gas turbine with waste-heat boiler energy-conversion subsystem for cogeneration application are shown in Figure 28. The ideal energy utilization factor and the gross process heat to electrical energy ratio are shown as a function of gas-turbine compressor pressure ratio and process-steam supply pressure (ideal process-heat temperature). The ideal energy utilization factor and the gross process heat to electric energy ratio are very weak fuctions of the expander inlet temperature, but rather strong functions of the process-steam supply pressure and compressor pressure ratio.

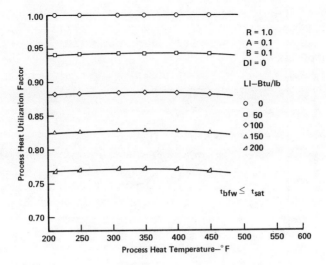

FIGURE 11. Characterization of process-heat utilization syb-
system for use with back-pressure turbine.

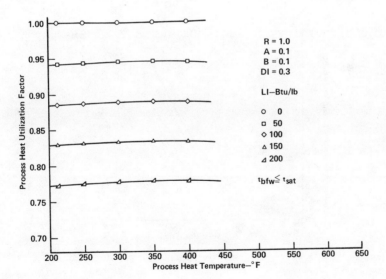

FIGURE 12. Characterization of process-heat utilization subsystem for use
with back-pressure turbine.

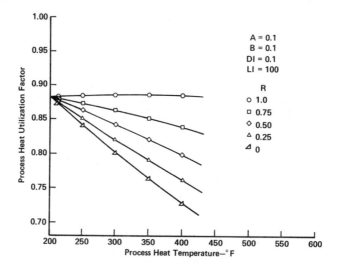

FIGURE 13.   Characterization of process-heat utilization subsystem for use with back-pressure turbine.

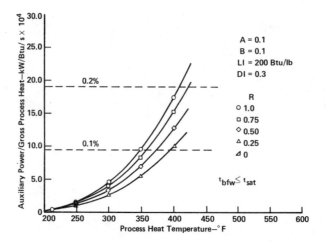

FIGURE 14.   Characterization of process-heat utilization subsystem for use with back-pressure turbine.

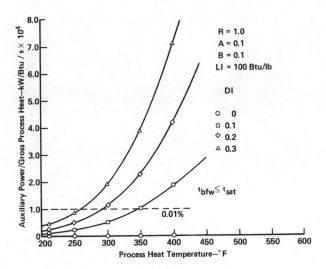

FIGURE 15.   Characterization of process-heat utilization subsystem for use with back-pressure turbine.

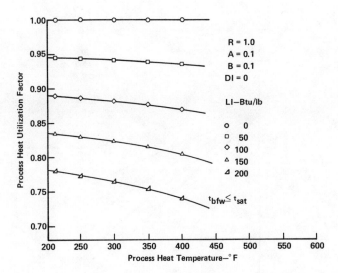

FIGURE 16.   Characterization of process-heat utilization subsystem for use with waste-heat boiler.

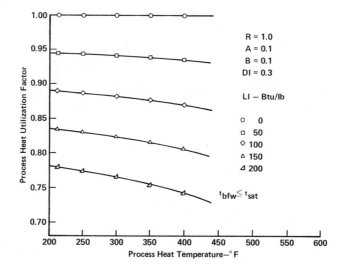

FIGURE 17.   Characterization of process-heat utilization subsystem for use with waste-heat boiler.

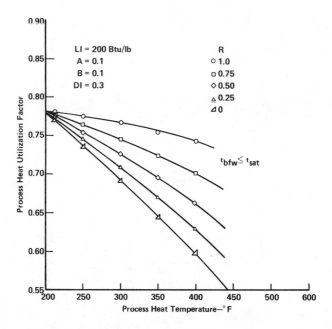

FIGURE 18.   Characterization of process-heat utilization subsystem for use with waste-heat boiler.

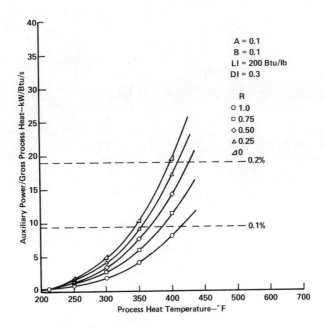

FIGURE 19. Characterization of process-heat utilization sub-system for use with waste-heat boiler.

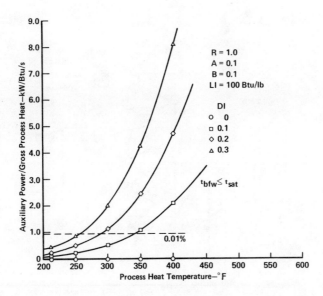

FIGURE 20. Characterization of process-heat utilization sub-system for use with waste-heat boiler.

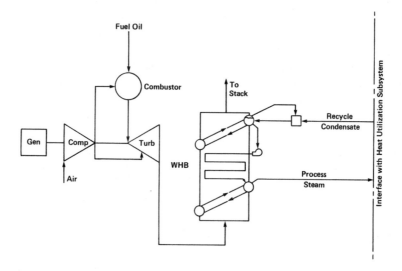

FIGURE 21.   Oil-fired gas-turbine energy conversion subsystem with waste-heat boiler.

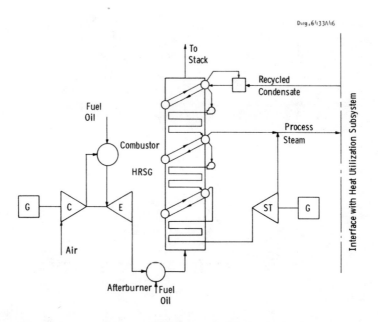

FIGURE 22.   Oil-fired combined-cycle energy conversion subsystem with modified heat-recovery steam generator.

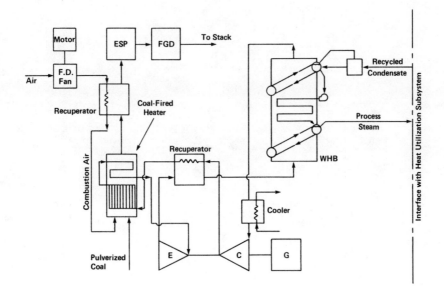

FIGURE 23. Closed-cycle gas-turbine energy conversion subsystem.

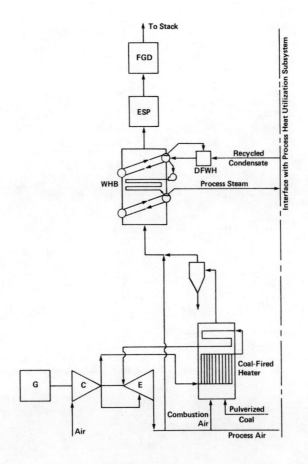

FIGURE 24. Coal-fired exhaust-heated energy conversion subsystem.

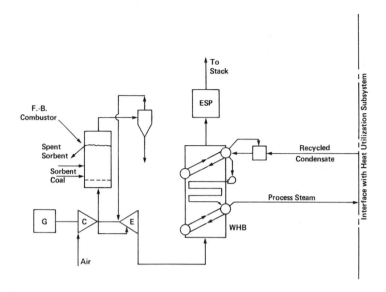

FIGURE 25.   Coal-fired gas-turbine energy conversion subsystem with fluidized-bed combustor and waste-heat boiler.

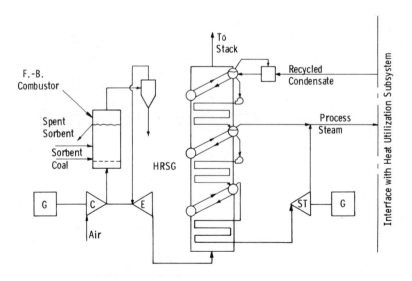

FIGURE 26.   Coal-fired combined-cycle energy conversion with fluidized-bed combustion and modified HRSG.

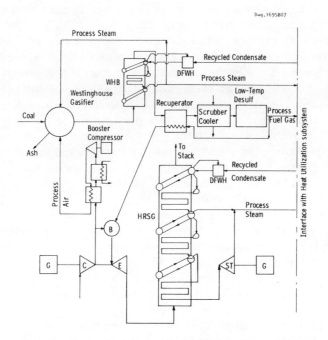

Dwg.1695807

FIGURE 27. Combined-cycle with integrated low-Btu gasification energy conversion subsystem.

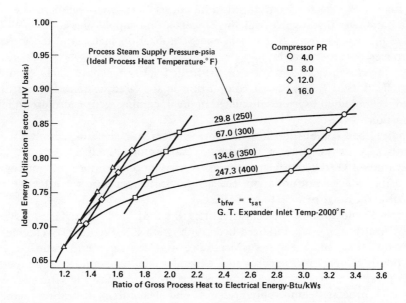

FIGURE 28. Characterization of oil-fired gas-turbine energy conversion subsystem with waste-heat boiler.

The characteristics of the oil-fired combined-cycle energy-conversion subsystem for cogeneration applications are shown in Figures 29 through 34. The ideal energy utilization factor and the gross process heat-to-electric energy ratio are shown as functions of gas-turbine compressor pressure ratio, gas-turbine expander inlet temperature, process-steam supply pressure (ideal process-heat temperature), and heat-recovery steam-generator inlet temperature. The energy utilization factor and the heat-to-electrical energy ratio are rather weak functions of the expander inlet temperature, but are fairly strong functions of the process-steam supply pressure and the compressor pressure ratio. The heat-recovery steam-generator inlet temperature has significant effects on both the energy utilization factor and the heat-to-electrical energy ratio.

There are many possible variations of the basic configuration shown in Figure 22. The use of an extraction steam turbine instead of a back-pressure steam turbine would provide a range in the process heat-to-electric energy ratio. Two levels of process-steam supply pressure could be provided by making the intermediate evaporator pressure either higher or lower than the steam turbine back pressure or extraction pressure.

The characteristics of a coal-fired closed-cycle gas-turbine energy-conversion subsystem (see Figure 23) are shown in Figure 35. The design values for gas-turbine compressor pressure ratio and expander inlet temperature which were used are those which have been used commercially in Europe.[17] Both the ideal energy utilization factor and the ratio of gross process heat-to-electrical energy are shown to be rather strong functions of the recuperator effectiveness. As has been the case in other subsystems, these parameters are also moderately strong functions of process-steam supply pressure.

The characteristics of the coal-fired exhaust-heated energy-conversion subsystem (see Figure 24) are shown in Figure 36 for a compressor pressure ratio of 4 and an expander inlet temperature of 1325°F. The state-of-the-art fired-heater technology from the closed-cycle gas turbine was assumed to be applicable to the exhaust-heated subsystem, although this configuration has never been used commercially. The only design parameter investigated was the process-steam supply pressure, which is shown to have the usual effect on the ideal energy utilization factor and the gross process heat-to-electrical energy ratio.

Both the closed-cycle and the exhaust-heated subsystems are likely candidates for the application of fluidized-bed combustion in the fired heaters. Several of the advanced technology subsystems discussed in the upcoming section also use fluidized-bed combustion.

Fluidized-bed combustion has the potential for in-bed desulfurization using solid sorbents such as dolomite or limestone. Coal-burning fluidized beds are limited to temperatures 100 to 200°F below the ash-softening temperature to avoid agglomeration. With in-bed desulfurization, the maximum temperature for effective sulfur removal is in the range of 1750 to 1850°F.

Figures 37 and 38 show the characteristics of a coal-fired gas-turbine subsystem with a waste-heat boiler using fluidized-bed combustion (see Figure 25). The ideal energy utilization factor and the gross process heat-to-electrical energy ratio are shown as functions of compressor pressure ratio, expander inlet temperature, and process-steam supply pressure for recycle condensate temperatures equal to the saturation value for the specific process-steam supply pressure. The ideal energy utilization factor and the gross process-heat-to-electrical energy ratio are rather weak functions of gas-turbine expander inlet temperature.

Figures 39 and 40 show the effect of recycle condensate subcooling on the subsystem characteristics over ranges of expander inlet temperature and compressor pressure ratio for a single value of process-steam supply pressure.

The combined-cycle energy conversion subsystem shown in Figure 26 is character-

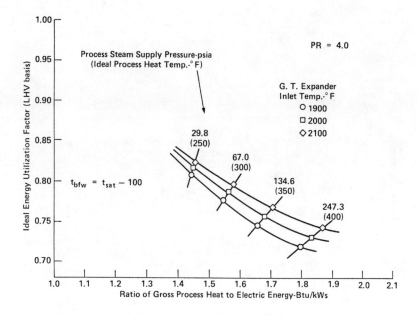

FIGURE 29. Characterization of oil-fired combined-cycle energy conversion subsystem with modified/unfired HRSG.

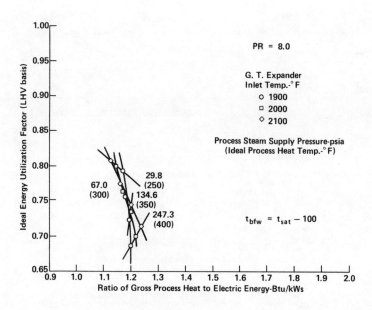

FIGURE 30. Characterization of oil-fired combined-cycle energy conversion subsystem with modified/unfired HRSG.

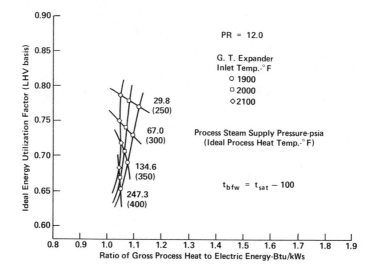

FIGURE 31.   Characterization of oil-fired combined-cycle energy conversion subsystem with modified/unfired HRSG.

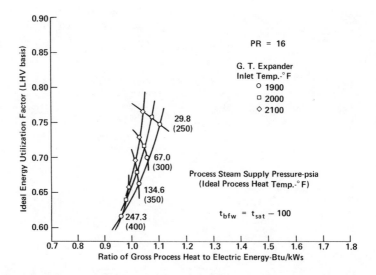

FIGURE 32.   Characterization of oil-fired combined-cycle energy conversion subsystem with modified/unfired HRSG.

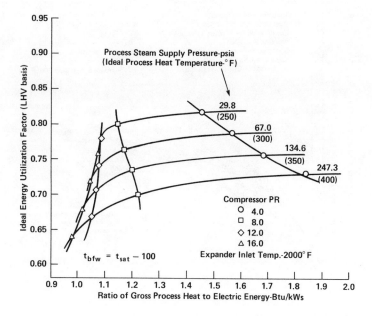

FIGURE 33. Characterization of oil-fired combined-cycle energy conversion subsystem with modified/unfired HRSG.

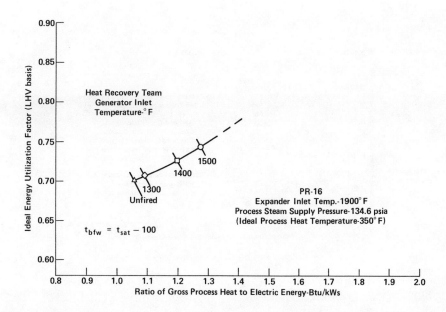

FIGURE 34. Characterization of oil-fired combined-cycle energy conversion subsystem with modified/fired HRSG.

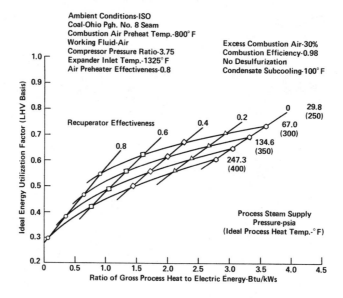

FIGURE 35.   Characterization of coal-fired closed-cycle gas-turbine energy conversion subsystem.

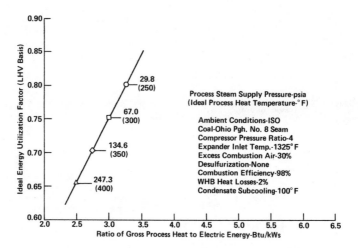

FIGURE 36.   Characterization of coal-fired exhaust-heated gas-turbine energy conversion subsystem.

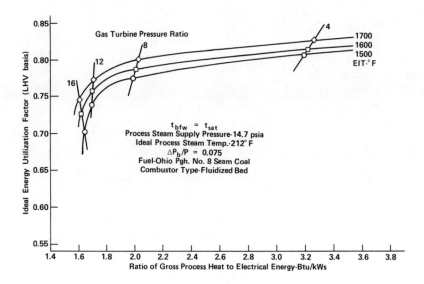

FIGURE 37.   Characterization of coal-fired gas-turbine energy conversion subsystem with waste-heat boiler.

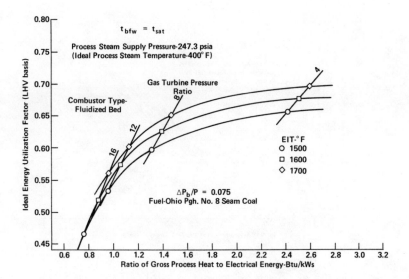

FIGURE 38.   Characterization of coal-fired gas-turbine energy conversion subsystem with waste-heat boiler.

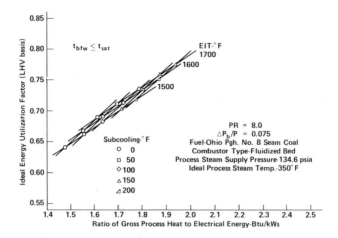

FIGURE 39.   Characterization of coal-fired gas-turbine energy conversion subsystem with waste-heat boiler.

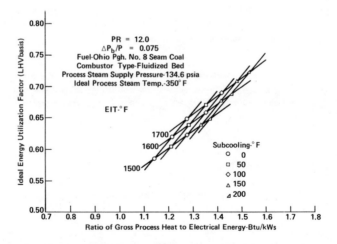

FIGURE 40.   Characterization of coal-fired gas-turbine energy conversion subsystem with waste-heat boiler.

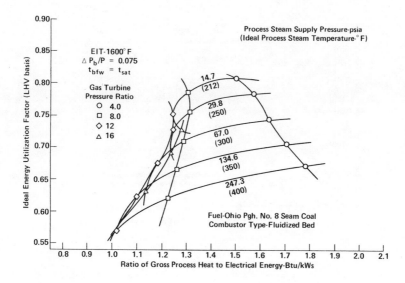

FIGURE 41. Characterization of coal-fired combined-cycle energy conversion subsystem with modified HRSG.

ized in Figure 41 as a function of compressor-pressure ratio and process-steam supply pressure, with the recycle condensate temperature equal to the saturation temperature for the specific process-steam supply pressure. Here, also, the ideal energy utilization factor and the gross process heat-to-electrical energy ratio are rather weak functions of gas-turbine expander inlet temperature. Figure 42 shows that the effect of recycle condensate subcooling is rather small for this energy conversion subsystem.

The combined-cycle energy conversion subsystem with the integrated low-Btu coal gasification shown in Figure 27 is characterized in Figure 43. In this subsystem, process fuel gas is supplied in addition to process heat and electrical energy. Figure 43 shows that both the ideal energy utilization factor and the gross process heat-to-electrical energy ratio are fairly strong functions of the fraction of product fuel gas produced in the gasifier which goes to process. As this fraction approaches 1, the ideal energy utilization factor becomes asymptotic to a value of approximately 0.78. This is so because the fuel gas to process is credited with 100% of its heating value in computing the ideal energy utilization factor.

A summary of the characteristics of the various energy-conversion subsystems using combustion turbines is given in Table 2.

III. DESIGN OF COGENERATION SYSTEMS

Cogeneration systems using combustion (gas) turbines can be designed for specific industrial applications by a synthesis of energy conversion- and process-heat utilization subsystems.

First, the process-heat utilization subsystem must be defined so that it can be characterized. This requires the following data:

- Process-heat temperature(s)
- Net process heat-to-electrical energy ratio
- Thermal losses from subsystem
- Pressure drops in subsystem lines
- Condensate losses or consumption
- Auxiliary power requirements of subsystem

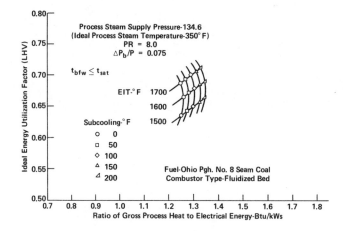

FIGURE 42.   Characterization of coal-fired combined-cycle energy conversion subsystem with modified HRSG.

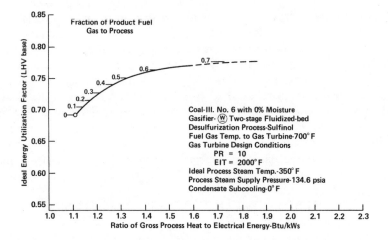

FIGURE 43.   Combined-cycle energy conversion subsystem with integrated low-Btu gasification, modified/HRSG, and waste-heat boiler for heat recovery from fuel gas.

TABLE 2

Summary of Combustion Turbine Energy-Conversion Subsystem Characteristics

Energy-conversion subsystem	Ideal process-heat temperature (°F)	Ratio of gross process-heat to electrical energy	Ideal energy utilization factor (LHV basis)
Oil-fired gas turbine with waste-heat boiler	250	1.55—3.35	0.77—0.87
	300	1.43—3.25	0.73—0.85
	350	1.28—3.10	0.69—0.82
	400	1.15—2.95	0.65—0.79
Oil-fired combined cycle with back-pressure steam turbine	250	1.15—1.45	0.75—0.82
	300	1.03—1.58	0.70—0.80
	350	1.00—1.70	0.66—0.77
	400	0.96—1.87	0.62—0.74
Coal-fired closed cycle with WHB	250	0.90—3.60	0.55—0.74
	300	0.60—3.30	0.47—0.70
	350	0.4—3.1	0.38—0.65
	400	0.1—2.8	0.30—0.61
Coal-fired exhaust-heated cycle with WHB	250	3.25	0.80
	300	3.0	0.75
	350	2.75	0.70
	400	2.50	0.65
Coal-fired gas turbine with WHB	250	1.5—3.15	0.60—0.80
	300	1.25—2.95	0.59—0.77
	350	1.0—2.75	0.53—0.73
	400	0.75—2.60	0.46—0.70
coal-fired combined cycle with back-pressure steam turbine	250	1.2—1.65	0.66—0.80
	300	1.1—1.7	0.60—0.76
	350	1.0—1.7	0.53—0.72
	400	0.95—1.8	0.54—0.69
Combined cycle with gasification, back-pressure steam turbine and WHB	350	1.1—2.0	0.69—0.78

Having the above listed data on the process-heat utilization subsystem will permit the computation of the process-steam supply pressure, the process-heat utilization factor, and consequently, the ratio of gross process heat-to-electrical energy. These parameters are required in order to select the energy-conversion subsystem configuration and design conditions. The ideal energy utilization factor for the energy-conversion subsystem is determined by the subsystem configuration and design conditions plus the process-steam supply pressure. Combining the heat utilization factor for the process-heat utilization subsystem and the ideal energy utilization factor for the energy-conversion subsystem gives the overall utilization factor for the cogeneration system.

Assume a plant with the following process heat and electrical energy requirements, for example:

- Process-heat temperature, 325°F
- Heat losses in steam supply line, 100 Btu/lb
- Heat losses in condensate return line, 10 Btu/lb
- Pressure loss in steam supply line, 30%
- Pressure loss in condensate recycle line, 3%
- Condensate recycle fraction, 100%
- Net process-heat requirement, 100×10^6 Btu/hr
- Electrical energy requirements, 10,000 kW
- Net process heat-to-electric energy ratio, 2.78 Btu/kWs

When the process steam supply is from a back-pressure or extraction turbine with an expansion line typical of industrial turbines, the heat utilization factor as obtained from Figure 12 is 0.89. When interfacing with energy-conversion subsystems that supply dry and saturated process steam, the heat utilization factor as obtained from Figure 17 is 0.88.

The pertinent data for energy-conversion subsystems that will interface with this heat utilization subsystem are as follows:

- Process-steam supply pressure, 135 psia
- Ideal process heat temperature, 350°F
- Recycled condensate subcooling, 35°F
- Gross process heat-to-electrical energy ratio for a back-pressure turbine = 2.78/0.89 = 3.12, for a waste-heat boiler = 2.78/0.88 = 3.16

Inspection of the characterization plots for the various energy-conversion subsystems treated herein shows that the following cases will nominally match the requirements of this heat utilization subsystem:

1. Oil-fired gas turbine with WHB with a compressor pressure ratio of 4.0 and an expander inlet temperature of 2100°F
2. Coal-fired closed cycle with a compressor pressure ratio of 3.75, an expander inlet temperature of 1325°F, and no recuperator

None of the other energy-conversion subsystems will exactly match the requirements for the assumed process-heat utilization subsystem. It is not necessary, however, that the output of the energy-conversion subsystem exactly match the requirements of the process-heat utilization subsystem. It can be sized to match the electrical energy requirements with an export of process steam, where there is an excess, or an import of process steam, where there is a deficiency. It can also be sized to match the process-steam requirements with an export or import of electrical energy. Since electrical energy is more easily transported than is process steam, the more likely choice would be to size the energy-conversion subsystem to match the process-steam requirements.

Two gas-turbine-based cogeneration systems that export electrical energy were evaluated. These are

- Coal-fired gas turbine with waste-heat boiler
- Coal-fired combined cycle with back-pressure steam turbine.

A comparison of the performance of four different cogeneration systems using gas turbines is shown in Table 3. In two cases, the energy-conversion subsystems are sized to meet exactly the process heat and the electrical energy requirements. In the other two cases, the energy-conversion subsystems are sized to meet the process-heat-requirement, and the excess electrical energy generated is exported to a utility.

For so-called Total Energy systems (such as Cases 1 and 2) where the process heat and electrical energy requirements are exactly matched, there is an apparent correlation between the overall energy utilization factor and the percent fuel savings. For those cases (3 and 4) where the process-heat requirements are matched and excess electrical energy is exported, there is no apparent correlation between the overall energy utilization factor and fuel savings. In these cases, the amount of excess electrical energy generated has a greater effect on the percent of fuel savings than the overall energy utilization factor.

Reference 21 gives costs for oil-fired gas turbine cogeneration systems with waste heat boilers.

Factor	Case			
	1	2	3	4
Gas turbine rating (kW)	10,150	24,050	60,000	73,200
Steam generated (lb/hr)	69,930	135,483	298,138	352,462
Process-steam supply pressure (psia)	165	165	165	165
Ideal process heat temp. (°F)	366	366	366	366
Gross process heat-to-electrical energy ratio (Btu/-kWs)	1.8	1.5	1.3	1.3
Capital costs ($000) (in 1977 dollars)				
Gas turbines and accessories	2,262.3	3,238.9	6,603.7	7,852.4
Heat-recovery steam generators	1,075.5	2,298.3	5,105.0	5,984.0
Stack	79.4	90.5	129.0	136.2
Deaerators	10.2	19.8	34.2	33.2
Boiler feed pumps and drivers	8.8	17.0	37.8	44.7
Fuel oil forwarding system	39.7	43.6	141.3	147.1
HRSG chemical treatment	5.4	11.5	25.5	29.9
Foundations	166.2	590.5	614.7	886.8
Earthwork and grading	24.9	42.9	81.6	93.3
Piping and valves	25.7	34.4	102.2	118.1
Ductwork and insulation	103.1	313.8	1,585.2	2,158.7
Control room	40.0	40.0	75.0	75.0
Oil storage tanks	197.5	422.2	937.6	1,099.2
Electrical cabling and switchgear	192.5	340.2	947.0	1,189.9
Sumps, culverts and revetments	220.0	278.7	371.5	389.8
Roads and paving	56.1	78.0	121.0	133.1
Land and land rights	152.0	262.0	498.0	570.0
Water treatment	218.2	422.7	930.2	1,099.7
Switchyard	126.9	300.6	750.0	915.0

	Case			
Factor	1	2	3	4
Total direct costs	5,004.4	8,845.6	19,090.5	22,056.1
Contingency @ 20%	1,000.9	1,769.1	3,818.1	4,591.2
Engineering and construction management @ 15%	750.7	1,326.8	2,863.6	3,443.4
Total indirect cost	1,751.5	3,096.0	6,681.7	8,034.6
Total gas turbine/WHB plant cost	6,755.9	11,941.6	25,772.2	30,990.7

From Burns & Roe, Inc.

Doherty[22] presents techniques for determining the profitability of industrial gas turbines that are applicable to gas-turbine-based energy-conversion subsystems for cogeneration applications. Three economic evaluation yardsticks are discussed: payout time, discounted cash flow, and total owning and operating costs, using fixed charges on investment.

TABLE 3

Comparison of Performance of Cogeneration Systems Using Gas Turbines Applied to Assumed Industrial Plant

Energy-conversion subsystem	Oil-fired gas turbine with WHB	Coal-fired closed-cycle gas turbine	Coal-fired gas turbine with WHB	Coal-fired combined cycle with back-pressure turbine
Gas turbine operating conditions				
Pressure ratio	4:1	4:1	4:1	4:1
Expander inlet temp. (°F)	2100	1325/No Recuperator	1700	1700
Ideal energy utilization factor (LHV basis)	0.79	0.62	0.748	0.722
Gross process heat-to-electrical energy ratio (Btu/kWs)	3.16	3.16	2.865	1.695
Gross electrical energy (kW)	10,000	10,000	11,030	18,407
Electrical energy exported (kW)	0	0	1030	8407
Net electrical energy (kW)	10,000	10,000	10,000	10,000
Overall energy utilization factor (LHV basis)	0.717	0.592	0.681	0.670
Fuel savings[a] (%)	22.2	5.8	19.2	24.5

[a] Based on the following assumptions:

1. The efficiency of a conventional coal-fired utility plant with FGD (LHV basis) is 35 ½%.
2. The transmission and distribution losses for electrical energy from the utility = 10%.
3. The efficiency of a coal-fired industrial boiler with FGD (LHV basis) is 85%.
4. The losses for utility-generated electricity and exported on-site generated electricity are equal.

LIST OF SYMBOLS

A   = ratio of heat loss in condensate return line to that in process-steam supply line.

B   = ratio of pressure loss in condensate return line to that in process-steam supply line.

D_l = fraction pressure loss in process-steam supply line.

EIT   = gas-turbine expander inlet temperature.

L1   = heat loss in process-steam supply line (Btu/lb).

PR   = gas-turbine compressor pressure ratio.

$\Delta P_b/P$ = ratio of combustor-loop pressure loss to combustor inlet pressure.

R   = condensate recycle fraction.

t_{bfw} = boiler feedwater temperature (°F).

t_{sat} = saturated steam temperature at process-steam supply pressure (°F).

GENERAL REFERENCES

Sawyer, J. W., Ed., Sawyer's Gas Turbine Engineering Handbook, Vol. 1, 2, 3, 2nd ed., Gas Turbine Publications, Inc., Stamford, Conn., 1972.

A. Combined Cycles

Sheldon, R. C. and McCone, T. D., Performance characteristics of combined steam-gas turbine cycles, *Proc. Am. Power Conf.*, Vol. 24, Illinois Institute of Technology, Chicago, 1962.

Ray, K. E., A 133,500 kW combined-cycle generating plant, *Heat Eng.*, 37(3), 1957.

George, T. H., The World's First Large Combined Cycle (Steam Turbine-Gas Turbine) Generating Unit: How Is It Doing? Am. Soc. Mech. Eng.—Inst. Electr. Electron. Eng. National Power Conference, American Society of Mechanical Engineers, New York, 1964.

Stout, J. B., Walsh, J. J., and Mellor, A. G., A large combined gas turbine-steam turbine generating unit, *Proc. Am. Power Conf.*, Vol. 24, Illinois Institute of Technology, Chicago, 1962, 404.

Foster-Pegg, R. W., Selection of the combined cycle, *Proc. Am. Power Conf.*, Vol. 25, Illinois Institute of Technology, Chicago, 1963, 267.

Hawley, C. F., Steam generator design, *Proc. Am. Power Conf.*, Vol. 25, Illinois Institute of Technology, Chicago, 1963, 284.

Cox, A. R., Henson, L. B., and Johnson, C. W., Operation of San Angelo power station combined steam and gas turbine cycle, *Proc. Am. Power Conf.*, Vol. 29, Illinois Institute of Technology, Chicago, 1967, 401.

Jones, R. W. and Schoults, A. C., Design and operating Experience with Gas Turbine Combined Cycle Units, ASME 71-GT-22, American Society of Mechanical Engineers, New York, 1971.

Blaskowski, H. J. and Singer, J. G., Gas turbine boiler applications, *Combustion*, 28(11), 1957.

Stewart, J. C. and Streich, H. J., The Design and Application of the Gas Turbine Heat Recovery Boiler, ASME-GT-38, American Society of Mechanical Engineers, New York, 1967.

Hambleton, W. V., General Design Considerations for Gas Turbine Waste Heat Steam Generators, ASME 68-GT-44, American Society of Mechanical Engineers, New York, 1968.

Smith, A. J. and Crabtree, L. C., Salt Grass — 300 MW Combined Cycle, Am. Soc. Mech. Eng. Gas Turbine Conf., Houston, March 1971.

B. Velox/Supercharged Boilers

Greco, L., A new Brown Boveri Velox boiler in a municipal district heating plant, *Brown Boveri Rev.*, 46, 371, 1959.

Daman, E. L. and Richardson, E. L., Economics of medium-sized supercharged power plants, *Proc. Am. Power Conf.*, Vol. 19, Illinois Institute of Technology, Chicago, 1957, 209.

White, A. O., The Combined Gas Turbine-Steam Turbine Cycle with Supercharged Boiler and Its Fuels, ASME 57-A-264, American Society of Mechanical Engineers, New York, 1957.

Gorzegno, W. P. and Zoschak, R. J., Supercharging the Once-Through Unit, ASME 64-PWR-15, American Society of Mechanical Engineers, New York, 1964.

Gorzegno, W. P. and Zoschak, R. J., The Supercharged Steam Generator — Some Aspects of Design and Pressure Level Selection, ASME 66-GT/CMC-68, American Society of Mechanical Engineers, New York, 1968.

Silberring, L., The supercharged steam generator, *Combustion,* 38(1), 32, 1966.

Bienz, J., Sharan, H., and Kikinis, A., Economics of Gas-side Supercharging of Steam Generators in Oil and Gas Firing, paper presented at the 8th World Energy Conf., Bucharest, June 23 to July 2, 1971.

Pirsh, E. A. and Sage, W. L., Combined Steam Turbine-Gas Turbine Supercharged Cycles Employing Coal Gasification, paper presented at the Am. Chem. Soc., Symp. on Coal Combustion, Toronto, Canada, May 26, 1970.

Beecher, D. T., Energy Conversion Alternatives Study, Phase I Final Report, Vol. 11, Westinghouse Research and Development, Pittsburgh, Feb. 12, 1976.

C. Closed Cycles

Taygun, F. and Schmidt, D., Today's Achievements with Conventional Closed-Cycle Gas Turbines and Their Future Aspects in the Nuclear Field, paper presented at the 16th ASME Int. Gas Turbine Conf., 1971, Houston.

Bammert, K. and Deuster, G., Layout and Present Status of the Closed-Cycle Helium Turbine Plant Oberhaser, ASME Paper 74-GT-132, American Society of Mechanical Engineers, New York, 1974.

Beecher, D. T., Energy Conversion Alternatives Study, Phase I Final Report, Vol. 6, Westinghouse Research and Development, Pittsburgh, Feb. 12, 1976.

Himmelblau, A. and Norton, J., Concept for Fluidized Bed Combustion of Consol Char using a Closed-Cycle Helium Power Plant with an Estimate of the Price of Electric Power, Contract No. E(49-18)-2201, Energy Resources, Co., Inc., Cambridge, Mass., for the Energy Research and Development Administration, Washington, D.C., 1976.

Buxman, J., Griepentrog, H., and Weber, D., Steam and Closed Gas Turbine Cycle, ASME Paper No. 77-GT-24, American Society of Mechanical Engineers, New York, 1977.

D. Total Energy Systems

Stocks, W. J. R., Gas turbines for total energy, *Electr. Rev.,* London, 185, 719, 1969.

Bjerklie, J. W., Small gas turbines for total energy systems, *Actual Specif. Eng.,* 26(2), G38, 1971.

Frei, D. H., Gas turbine as a total energy system for paper mills, *Pulp Pap. Int.,* 16(10), 48, 1974.

Wittner, B. R. and Culp, R. E., Offshore application of gas-turbine-powered prepackaged total energy systems, *Proc. Am. Power Conf.,* Vol. 35, Illinois Institute of Technology, Chicago, 1973, 779.

Green, L., Jr., Gas-Turbine Total Energy System for a Cement Plant, MITRE Corp., McLean, Vã., ASME Paper No. 77-GI-22, American Society of Mechanical Engineers, New York, 1977.

Anon., Total energy systems and the gas turbine combined cycle, *Electr. Consult.,* 91(3), 24, 1975.

Hundemann, A. S., Total Energy Systems for Buildings, NTIS/PS-75/215, NTIS/PS-66/0275, National Technical Information Service, Springfield, Va., April 1976, 51.

Tintori, J., Schiefer, R. B., and Taylor, J. R., Fuel for Total Energy, Campagnei Industrielle Pont-de-Claix, Grenoble, France, ASME Paper No. 71-GT-55, American Society of Mechanical Engineers, New York, 1971.

Stocks, W. J. R., Gas turbine offers short cut to energy economy, *Energy Int.,* 11(4), 21, 1974.

REFERENCES

1. **Beecher, D. T.,** Energy Conversion Alternatives Study, Westinghouse Phase I Final Report, Vol. 4, Contract No. NA53-194D7, Westinghouse Research and Development, Pittsburgh, Feb. 12, 1976, 5-1.

2. **Strong, R. E.,** High Temperature Technology Program, Phase I — Program and System Definition, Contract No. EX-76-C-01-2290, prepared by Generation Systems Division, Westinghouse Electric Corporation, for U.S. Energy Research and Development Administration, May 1977.

3. **Noack, W. G.,** The Velox boiler, *Engrg.,* 135(3496), 52, 1933.

4. **Daman, E. L. and Zoschak, R. J.,** Supercharged boiler design, development, and application, *Proc. Am. Power Conf.,* Vol. 18, Illinois Institute of Technology, Chicago, 1956, 117.

5. **Mordell, D. L. and Foster-Pegg, R. W.,** Test of an experimental coal-burning turbine, *Trans. Am. Soc. Mech. Eng.,* 78, 1807, 1956.

6. **Fraas, A. P.,** Concept Preliminary Evaluation of Small Coal-Burning Gas Turbine for Modular Integrated Utility System, OCR R&D Rep. No. 96, Oak Ridge National Laboratory, Department of Housing and Urban Development (HUD), and Office of Coal Research (OCR), Department of Commerce, Washington, D.C., 1976.

7. **Moskowitz, S. and With, G.,** Pressurized fluidized bed pilot plant for production of electric power using high sulfur coal, *Proc. 12th Intersoc. Energy Conversion Engineering Conf.,* Vol. 1, American Nuclear Society, LaGrange Park, Ill., 1977, 696.

8. **Schuster, Ray,** The growing presence of gas turbine combined cycles, *Power Eng.,* 76(1), 28, 1972.

9. **Lecansi, A. C.,** The Supercharged Steam Generator: Its First Shipboard Installation, ASME Paper No. 62-WA-291, American Society of Mechanical Engineers, New York, 1962.

10. **Stephens, J. O.,** Gas turbines for blast-furnace blowing, *Mech. Eng.,* 80, 103, 1957.

11. **Yellott, J. I. and Kottcamp, C. F.,** The Coal-Fired Gas Turbine Power Plant, Locomotive Development Committee, Bituminous Coal Research, Inc., Monroeville, Pa., presented at the Am. Soc. Mech. Eng. Semi-Annual Meeting, Chicago, June 19, 1947.

12. Staff, Energy Conversion from Coal Utilizing CPU-400 Technology, Final Report, ERDA Contract No. EX-76-C-01-1536, Combustion Power Co., Menlo Park, Calif., for the Energy Research and Development Administration, Washington, D.C., 1977.

13. Staff, Combustion Power Unit-400, Contract No. Ph 86-67-259 by Combustion Power Co., Menlo Park, Calif., for the Bureau of Solid Waste Management, Rockville, Md., 1969.

14. **Pillsbury, P. W. and Lin, S S.,** Recent Tests of Industrial Gas Turbine Combustors Fueled with Simulated Low Heating Value Coal Gas, ASME Paper No. 76-WA/GT-3, American Society of Mechanical Engineers, New York, 1976.

15. **Cabal, A. V., Dabkowski, M. J., Heck, R. H., Stein, T. R., Chamberlin, R. M., Mulik, P. R., Singh, P. P., and Rovesti, W. C.,** Utilization of Coal-derived Liquid Fuels in a Combustion Turbine Engine, Vol. 23, No. 1, paper presented at the 175th Am. Chem. Soc. Nat. Meet., American Chemical Society, Washington, D.C., 1978.

16. **Congiu, A.,** A 37/42 MW Gas Turbine for Power Generation, ASME Paper No. 64-GTP-4, American Society of Mechanical Engineers, New York, 1964.

17. **Bammert, K. and Groschup, G.,** Status Report on Closed-Cycle Power Plants in the Federal Republic of Germany, ASME Paper No. 76-GT-54, American Society of Mechanical Engineers, New York, 1976.

18. **Meyer, A.,** High pressure boilers; Velox steam generator, *Mech. Eng.,* 40(9), 341, 1935.

19. **Bund, K., Henney, K.-A., and Krieb, K. H.,** Combined Gas/Steam-Turbine Generating Plant with Bituminous-Coal High Pressure Gasification Plant in the Kellerman Power Station at Lunen, 8th World Energy Conference, Bucharest, June 23 to July 2, 1971.

20. **McCay, F. L.,** Committee Chairman, The Coal-Burning Gas Turbine Project, Report of the Interdepartmental Gas Turbine Steering Committee, Department of Minerals and Energy, Department of Supply, Australian Government Publishing Service, Canberra, 1973.

21. **Bos, P.,** Project Director, The Potential for Cogeneration Development in Six Major Industries by 1985, Contract No. CR-04-60172-00, Resource Planning Associates, Inc., for the Office of Conservation and Solar Applications, Department of Energy, Cambridge, Mass., December 1977.

22. **Doherty, M. C.,** Investment, Economics of Industrial Gas Turbines, ASME Paper No. 77-GT-26, American Society of Mechanical Engineers, New York, 1977.

Chapter 8

DIESEL ENGINES WITH HEAT RECOVERY

J. P. Davis

TABLE OF CONTENTS

I. INTRODUCTION

The diesel engine concept was described by Dr. Rudolph Diesel in his dissertation *The Rational Heat Motor* in the 1890s. An important feature of Diesel's engine was the introduction of the fuel into the cylinder at the end of the compression stroke at which point the air had become hot enough through compression to ignite the fuel spontaneously, rather than having to depend on a "hot spot" as other designs required. His original aim was to combust the fuel at constant temperature in order to approach Carnot Efficiency as closely as possible. He didn't achieve this goal, but rather, came closer to constant pressure combustion than constant temperature combustion. His failure to achieve a Carnot Cycle resulted in the cycle of his own name.

In order to achieve the necessary spontaneous ignition temperature of the fuel, Diesel used considerably higher compression ratios than previously attempted in other heat engines. This resulted in a relatively heavy high pressure machine. The advantage, however, was notable improvement in thermal efficiency over other heat engines of the day.

An excellent article on the history and basic thermodynamics of Diesel's invention is contained in Reference 1.

II. CURRENT DIESEL ENGINE TYPES

Currently manufactured diesel engines can generally be categorized as follows:

Engine type	Speed range	Power range (MW)	Cycle	Typical shaft efficiency (LHV)
High speed	>900 rpm	1	4-stroke	34
Medium speed	400—900 rpm	1—10	4-stroke 2-stroke	38
Slow speed	<200 rpm	5—40	2-stroke	40

High-speed engines are generally restricted to operation on distillate fuels. Medium-speed engines may operate on both distillate and residual fuels, but U.S. practice has primarily been to restrict operation to distillates. In the future years, one would expect many more residual fuel applications in the medium-speed range if the current ratio of distillate/residual price is maintained or increases above its current level of 1.25 to 1.5. Slow-speed engines of the valveless two-stroke type have found widespread application in the marine propulsion field where they are the dominant diesel engine type at power levels beyond ~10,000 hp. These engines are large and relatively costly compared to other types. Their primary virtues are high efficiency, high reliability, and ability to digest a steady diet of poor-quality fuel. Typical fuel specifications for the various engine types are shown in Table 1.

III. MEDIUM-SPEED DIESEL ENGINE CHARACTERISTICS

One of the major diesel engine manufacturers in the U.S. is the Fairbanks-Morse Engine Division of Colt Industries. Characteristics of two basic diesel engines manufactured by Fairbanks-Morse are presented in Table 2. The detailed heat balance for the two engines are given in Figures 1 and 2. The heat balance data were provided by Fairbanks-Morse based on testing of the engines in their laboratories. In this testing, a back pressure of 10 to 12 in. water is normally placed on the engine.[2]

TABLE 1

Residual Oil Fuel Specifications for Diesel Engines

	Engine speed		
	120 rpm	514 rpm	900 rpm
Specific gravity at 15°C (g/cm³)	0.990	0.98	0.940
Viscosity at 100°F (seconds red-wood I)	6000	3500	1000
Sulfur (wt %)	5.0	4.0	3.0
Conradson (wt %)	15.0	10.0	8.0
Vanadium (ppm)	300	200	100
Sodium (ppm)	1/3	65	1/3
H_2O (vol %)	<2.0	2.0	<0.5
Cetane number	>25	>40	>40
Flash point (i.h.T.°C)	>65	>65	>65

TABLE 2

Characteristics of Selected Diesel Engines

Model	Colt 38TD8-1/8			Colt-Pielstick PC-2					
Type	Opposed piston, turbocharged, two cycle			V-configuration, turbocharged, four cycle					
Speed, rpm	900			514					
hp/Cylinder									
Base rated (100%)	300			500					
Peak rated (110%)[a]	330			550					
BMEP, base rated, psi	127			219					
Bore, in.	8 1/8			15.75					
Piston stroke, in.	10.0			18.11					
Displacement per piston, in.³	518			3528					
Compression ratio	13.8/1			13/1					
Number of cylinders/engine	6	9	12	8	10	12	14	16	18
Base-rated hp	1800	2700	3600	4000	5000	6000	7000	8000	9000
Fuel capability	No. 2 diesel fuel, dual fuel[b]			No. 2 diesel fuel. residual and other heavy fuels, dual fuel					

[a] Overload of 10% permitted for 2 hr out of 24.
[b] Gas with compression ignition of pilot oil.

IV. SLOW-SPEED DIESEL ENGINE CHARACTERISTICS

The large, slow-speed diesel engine with residual oil burning capability was originally developed for marine applications. The slow-speed (100 to 150 rpm) is desired to permit direct coupling with the propeller and thereby eliminate the need for massive reduction gears in the propulsion system. The slow-speed engine is also compatible with the longer burning times required for heavy marine fuels such as Bunker "C" or residual. To achieve reliable operation with these heavy fuels, the engine design is based on the two-stroke cycle to eliminate valves and on the use of a crosshead piston to prevent contamination of the crankshaft lubricant with the residual oil combustion products. The crosshead constrains the power piston connecting rod to a linear movement. A seal surrounding the rod effectively isolates the upper fuel burning section of the engine from the lower crankcase. To increase the power density of the engine, the inlet air is compressed by a turbocharger operated from the engine exhaust gases.

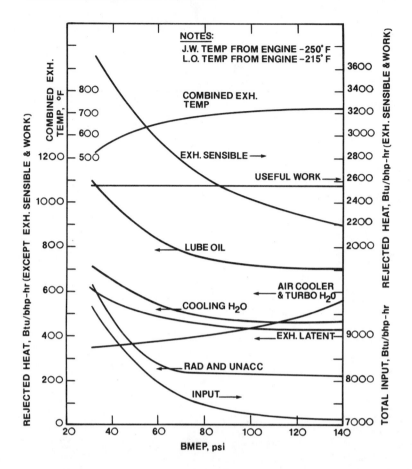

FIGURE 1.   Thermal balance, Model 38TD8-1/8 and 38TDD8-1/8 oil-Diesel operation at 900 rpm. (From Colt Industries, F. M. Engine Division, February 22, 1974. With permission.)

Engines of this type have dominated the marine propulsion field since World War II. The engines are manufactured with multiple cylinders driving a single crankshaft to meet various power demands. Typical engines contain 6 to 12 cylinders with the power per cylinder available up to 4000 hp. Large 48,000 hp engines have been in service for 10 years.

The same basic engine design is used for stationary land applications. Most of these engines are directly coupled to generators for electrical power production. These plants are located in many countries in Europe, Africa, and the Far East. As a notable exception, none of these large units are found in the United States. In this country, the traditional use of steam turbine technology has effectively prevented the development and use of large diesel-electric systems.

Large diesel engines are manufactured presently only by foreign companies. These include Sulzer Brothers Ltd., Switzerland; MAN, Germany; Burmeister and Wain, Denmark; Mitsubishi, Japan; and Grandi Motori Trieste, Italy. Sulzer is the largest manufacturer. Sulzer engines and parts are presently being built under license in 17 countries. The engines are manufactured to the original drawings, and all parts are interchangeable regardless of the origin of manufacture. Sulzer has recently licensed Westinghouse to build the engines in the U.S. Thus, a domestic supplier will soon be available.

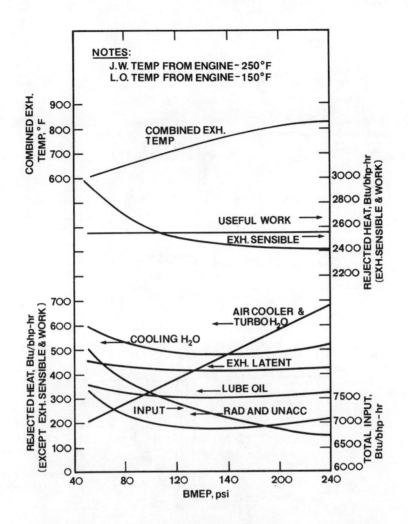

FIGURE 2. Thermal Balance, Model PC-2, oil-Diesel operation at 500 rpm. (From Colt Industries, F. M. Engine Division, February 22, 1974. With permission.)

Future developments in diesel technology include the possible use of coal-derived fuels. The present dual fuel (gas-oil) capability, and the past experience in burning pulverized coal, strongly suggests that modifications to the present engine design will permit the burning of a variety of coal-derived fuels. Fuels of this type are presently being developed to encourage the greater use of coal and, thereby, reduce our dependence on petroleum.

A cross section of the large two-stroke diesel engine with an output power of approximately 20,000 metric hp* is shown in Figure 3. The engine has six in-line cylinders and operates at 120 rpm at rated power.

These engines are cooled by a closed forced-circulation water-cooling system. Individual circuits provide cooling for the lubricating oil, the pistons, the jacket (cylinder head and walls), and the charge air (after compression in the turbocharger). Typical heat loads for these components are as follows:

* metric hp = 0.9863 U.S. hp

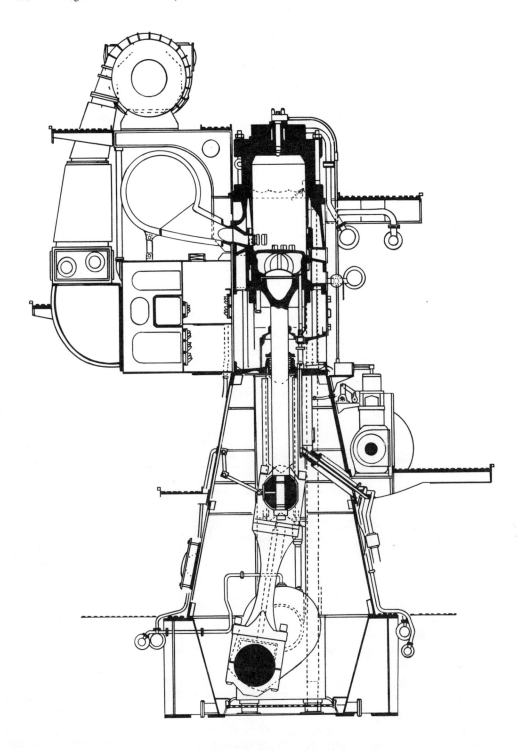

FIGURE 3.   Sulzer Diesel engine cross section.

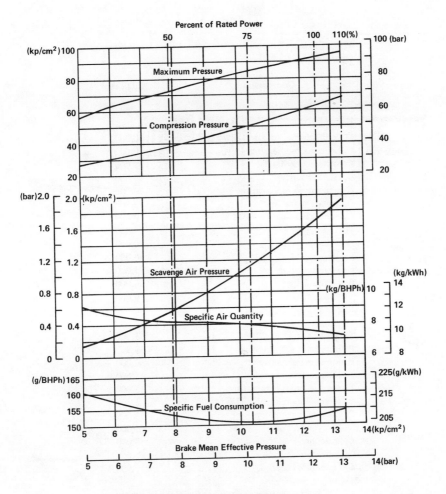

FIGURE 4.   Diesel engine performance.

Component	Value
Oil coolers	10 kcal/hph**
Piston coolers	43 kcal/hph
Jacket coolers	142 kcal/hph
Charge air coolers	160 kcal/hph

The engine is started using compressed air. Fuel oil is supplied by a fuel preparation system that removes water and solid impurities and steam heats the fuel to the required viscosity level.

Figure 4 illustrates the engine performance as a function of power output. Specific fuel consumption is relatively constant over a wide power range from 50 to 110% of rated load. At full load, fuel consumption is 152 g/hph. This reduces to about 150 g/hph at three quarters load, and 153 g/hph at one half load. Engine speed is automatically regulated by a governor. System control stations can be located either remotely or adjacent to the engine. In marine applications, the engine rooms are typically unattended except for brief inspections by one person. It is general practice to have no-watch periods of 16 hr on weekdays and 44 hr on weekends (the engine room is unattended for 120 hr out of the 168 hr during the week).

** metric hp - hr

TABLE 3

Main Engine Components Working Life Heavy-Fuel-Oil Operation

Component	Useful life (hr)
Crankshaft	no limit
Crosshead pin	no limit
Main bearing	50,000
Connecting rod bearing	50,000
Crosshead bearing	35,000
Crosshead slipper	60,000
Crosshead oil-link bearing	70,000
Camshaft gear drive[a]	70,000
Camshaft bearing[a]	70,000
Fuel cam, plunger[a]	35,000
Fuel pump valve[a]	20,000
Fuel nozzles[a]	10,000
Governor	35,000
Governor drive	70,000
Turning gear	90,000
Starting air control valve	60,000
Starting air inlet valve	30,000
Scavenge air valves	25,000
Piston head[a]	35,000
Piston skirt[a]	50,000
Piston ring[a]	12,000
Piston rod stuffing box. rings[a]	15,000
Piston cooling telescopic gland[a]	6,000
Piston cooling telescopic pipes[a]	35,000
Outer cylinder cover[a]	50,000
Cylinder head[a]	50,000
Cylinder liners[a]	45,000
Cylinder lubricators	50,000

[a] The above represent the lower limits when running under particularly unfavorable conditions. For operation on gas, add one third to the life of components.

For electrical power generation, the engine is directly coupled to a synchronous generator. The generator is equipped with a static exciter and automatic voltage regulation. The efficiency of the generator is typically 96%. Combining the generator efficiency with the auxiliary loads required for engine operation (water and fuel pumps, etc.) the net usable electrical power output is approximately 94% of the gross shaft power developed by the engine.

Maintenance requirements for the large slow-speed diesels are significantly reduced compared with higher-speed engines. Sulzer's experience on the life of parts is summarized in Table 3 for heavy-oil operation. The majority of parts have a useful life of 50,000 hr or longer. The parts that require replacement after 10,000 hr or less are the fuel nozzles and the sealing glands for the piston cooling circuit. Parts with the shortest life are the least expensive. The recommended downtime for engine inspection and maintenance is 300 to 450 hr/year. This downtime is equivalent to a plant availability of about 95%.

The exhaust emissions from large slow-speed diesel engines are similar to those emitted from diesel-powered locomotives and heavy-duty vehicles. Representative emission factors (lb of pollutant per 10^6 Btu fuel input) are listed in Table 4. The stationary

TABLE 4

Emission Factors (lbs/10^6 Btu Input)

	Utility powerplant[a] (residual oil)	Diesel locomotive, Two-stroke, turbocharged[a]	Heavy-duty Diesel trucks and buses[a]	Large stationary Diesel engine, slow-speed (residual oil)[b]
Particulates	0.053	0.17	0.087	0.02—0.06
SO$_2$	1.05[c]	0.95[c]	0.90[c]	1.05[c]
CO	0.02	1.07	1.5	0.11—0.59
HC	0.013	0.19	0.25	0.11—0.30
NO$_2$	0.3—0.7	2.2	2.5	2.2 —2.8

[a] From the Compilation of Air Pollutant Emission Factors, EPA, AP-42, 2nd ed. April 1973.
[b] Sulzer Bros., Ltd., Switzerland.
[c] Sulfur in fuel (wt %).

diesel engines tend to be lower in both particulates and carbon monoxide. Sulfur dioxide emissions are dependent on the amount of sulfur present in the fuel.

The possibility of burning coal-derived fuels (pulverized coal, coal-oil slurries, etc.) in large slow-speed diesel engines greatly expands the fuels available for industrial co-generation applications. The use of coal-derived fuels would reduce the overall national use of petroleum and would make maximum use of existing diesel engine technology with its demonstrated advantages of high electrical conversion efficiency and short construction times. Thermo Electron believes this coal-derived fuel capability can be developed by a moderate research and development effort and has proposed such work to the Office of Fossil Energy, Energy Research and Development Administration.

Engineers have worked on the problem of burning solid fuels, particularly coal, in combustion engines from the 18th century up to the present time. The earliest attempts seem to have been in France with the work of Joseph Montgolfier, and slightly later by the Niepce cousins whose pyreolophore was fired with power coal. Sadi Carnot was familiar with these early experiments, and his thermodynamic analysis of the pyreolophore* is contained in the classic *Reflections on the Motive Power of Fire* published in 1824.[3] It is interesting to note that Carnot's suggested improvements included a recommendation that the exhaust from the engine be used to generate steam in a boiler, thereby increasing the net work output from the system. This is the first known reference to the "bottoming cycle" concept.

In 1893, Rudolf Diesel arranged with Augsburg and Krupp of Germany to develop a more efficient internal combustion engine. One of Diesel's objectives was to use powdered coal as a fuel for his new engine.[4] The first engine was built that same year and was to have operated on powdered coal. Unfortunately, the engine exploded at the first attempt to start it, and all subsequent attempts to operate the engine using powdered coal ended in failure. Oil was finally adopted as the fuel.

Rudolf Pawlikowski was an associate of Diesel during this period, and evidently he was not discouraged by the early attempts to operate an engine with powdered coal. Beginning in 1911, and continuing for at least 15 years, Pawlikowski, with the support of the Kosmos Machine Works of Coerlitz, Germany, experimented with diesel engines converted to operate on powdered coal. His work appears to have been successful,

---

* The pyreolophore operated on the Lenoir Cycle and was designed to propel a boat. Air at atmospheric pressure was drawn into a cylinder, the powered fuel was introduced and ignited, and the resultant expansion of the air produced the power.

with cycle efficiency and wear rates about the same as oil engines of that time.[5-8] Subsequent efforts have been almost inconsequential when compared with Pawlikowski's work.

In the late 1930s, the Fuel Research Center in England conducted a series of tests as a preliminary to operating a diesel engine on coal dust.[9] These tests were to determine the best combination of materials to use for piston, rings, and cylinder liner and were carried out in a single-cylinder gasoline engine. Coal ash was injected into the carburetor, and wear rates of the various components were measured. World War II ended this work before an engine was actually operated on powdered coal. Since World War II, there have been a few attempts to burn coal in diesel engines in the U.S., culminating in the work sponsored by the Office of Coal Research between 1966 and 1969 at Howard University.[10] All of the post-war work falls into two categories: (1) either pulverized coal was mixed with diesel fuel, and the resultant colloidal fuel injected in the conventional manner; or (2) the pulverized coal was aspirated with the intake air, and diesel oil was injected in the conventional manner in order to effect ignition. All of this more recent work must be considered unsuccessful, as the wear rates and cycle efficiencies were completely unacceptable by modern standards.

At this point, it is worth examining the various attempts to burn pulverized coal in piston engines in more detail so as to determine why Pawlikowski's engine was relatively successful from the technical point of view. The Rupa engine (as Pawlikowski's engine was known) differed from Diesel's coal engine in its method of injecting the pulverized coal. Diesel used compressed air at a few hundred psi above compression pressure to blast a charge of coal into the cylinder through a cam-operated poppet valve timed to open near the end of the compression stroke. The fact that high temperature and pressure-combustion gases could get back into the coal and compressed air feed system was the most likely cause of Diesel's explosion problem. The Rupa coal-feed system operated at a low pressure and deposited a charge from a constantly circulating coal-air slurry into a kind of precombustion chamber during the suction stroke of the four-stroke engine. The valve that admitted the fuel charge also vented the precombustion chamber to atmosphere before allowing a new fuel charge in. At the end of the compression stroke, a small quantity of oil was also injected into the precombustion chamber, and a second valve operated to admit compressed air which blasted the entire charge into the main chamber. Evidently, the oil was added to ensure ignition, although this appears not to have been essential.

The success of the Rupa engine relative to later attempts is easily explained. With respect to wear rates, Pawlikowski took extreme precaution to keep the coal ash residue away from the cylinder liner and rings. He used compressed air jets in the liner to keep ash away from the rings and increased the lubricating oil consumption to wash the ash from the cylinder walls. He undoubtedly was aided by the 16.5 in. bore of his test engine. That he was successful is attested to by the conservative estimate of 9,000 hr of running time he was able to accomplish over a period of 12 years with wear rates not much above those of conventional oil engines of that period. As already mentioned, all subsequent efforts either attempted to aspirate the coal with the intake air, thereby depositing a significant amount on liner and rings, or mixed the coal with diesel oil and attempted to inject it through conventional injection equipment. It is not surprising that excessive wear was the result of both of these methods.

With respect to efficiency, the superiority of the Rupa engine stemmed primarily from its low operating speed (160 rpm). All the engines tested in the United States since World War II were operated at 800 rpm and higher, and cycle efficiency was uniformly poor. This was simply due to a lack of time for combustion of the coal particles. The engine tested at Value Engineering Company under subcontract from

Howard University was operated at 1000 rpm, and under some conditions the particulate matter in the exhaust was found to have a heating value about 90% that of the coal fuel.

The Rupa engine appears to have had a cycle efficiency of about 32%, and the brake mean effective pressure, 113 psi, was quite respectable for that time. The coal was pulverized to a point where 80% would pass through a 100-mesh screen, which is the approximate particle size used in modern coal-fired steam plants. The engine was operated using a wide variety of different rank coals including a mixture of 80% hard Silesian coal, containing 16% ash and 20% lignite. Even pulverized peat, sawdust, charcoal, rice dust, flour, and coke were used successfully as fuels for the Rupa engine.

Most of the published data on the Rupa engine are for a 16.5 in. bore by 25 in. stroke single-cylinder M.A.N. engine which had begun life as a diesel in 1906. This is the engine which accumulated the 9,000 hr of operating time and was still in commercial operation in 1928, supplying power for factory machinery. A few other engines were converted to operate on pulverized coal, including a 12.6 in. bore by 20.45 in. stroke three-cylinder engine that operated at 215 rpm. Worthy of note is the claim made for the Rupa engine that "... the exhaust is free of combustible matter, and is light brown in color. It contains no trace of tar, and during the course of several years there has been no blackening of a neighboring wall by the exhaust gases."

The reason that the Rupa engines were never widely accepted remains obscure. A possibility is that with the rapid advances in diesel technology occurring at that time the operating speed of the largest bore engines was probably increasing out of the range where the combustion of pulverized coal could be accomplished efficiently, and the development of the very large-bore slow-speed marine engines had not yet begun.

V. COGENERATION APPLICATIONS

The classic "total energy system" of past years often consisted of diesel engine(s) and heat-recovery boilers to generate steam and/or hot water for commercial applications, primarily buildings. Several circumstances led to their lack of widespread acceptance: (1) less than ideal design, (2) lack of experienced operating personnel, (3) highly variable power/thermal requirements, and (4) customers' primary concern with first costs rather than long-term operational savings. For industrial applications, these negative circumstances could be alleviated to a considerable degree.

Diesel topping should be well-suited to applications in the lower range of power/steam requirements from as low as 1000 kW of power requirement (corresponding to as little as 3000 lb/hr of steam) to beyond 30 MW. Although basic engine efficiencies are high ($\sim$35 to 40%), a substantial fraction of the reject heat is at too low a temperature for any recovery, e.g., lube oil heating. Cylinder heat losses could be used to generate low pressure steam at $\sim$15 psig, useful for some industrial applications, but not all. Heat rates are in the 6000 to 7000 Btu/kWh range (57% to 49% efficiency), not as favorable as steam or gas turbines. However, again, this is better than the typical utility heat rates.

The larger engines of the two-stroke marine propulsion diesel type are capable of routinely burning residual fuels not generally suitable for gas turbine use. A program may be initiated in the near future to investigate the possibility of firing pulverized coal and coal-derived liquid fuels in diesel engines.

Typical heat rates and power-to-steam ratios are shown in Figures 5 and 6, which also show these values for back-pressure steam turbines and gas turbines. Note that diesels are particularly suited to a high ratio of electric power to steam compared to the other systems.[11]

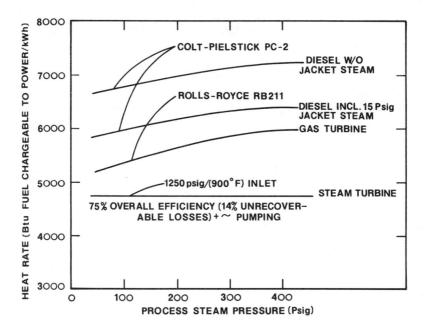

FIGURE 5.   Heat rates vs. steam pressure.

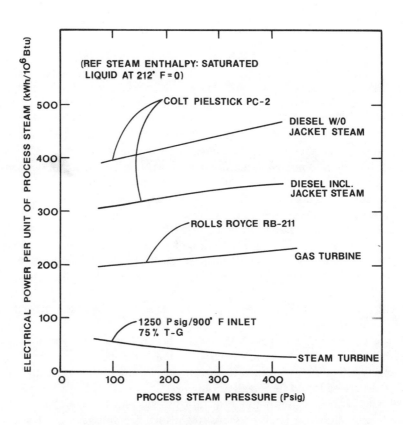

FIGURE 6.   Electric power generation capabilities per unit of process steam.

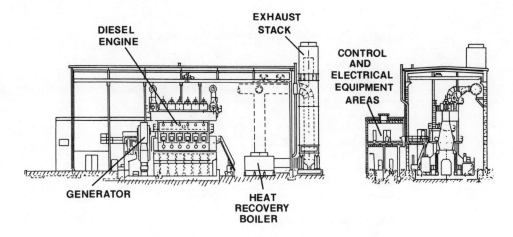

FIGURE 7.   Proposed Diesel cogeneration system.

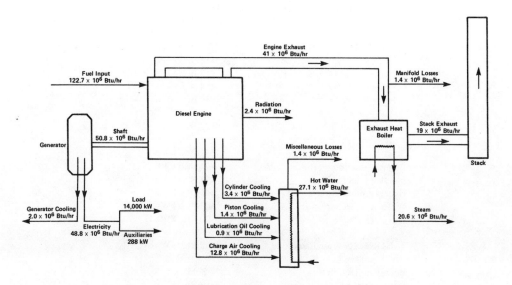

FIGURE 8.   Diesel cogeneration system (14,000 kW).

VI. ECONOMICS OF A TYPICAL INDUSTRIAL APPLICATION

A conceptual view of the proposed diesel cogeneration system is shown in Figure 7. The diesel engine, generator, exhaust-heat recovery boiler, and engine subsystems are housed in a high-bay building. A diverting valve in the exhaust system passes the exhaust to the heat-recovery boiler or, if required, directly to the adjoining stack. The stack is equipped with a silencer or muffler to reduce the exhaust noise to acceptable levels. The building contains an overhead crane with a capacity sufficient to carry out all required maintenance. The engine subsystems include water cooling, lubrication, heavy-fuel-oil treatment, and control and starting equipment. A building addition to the side of the main high-bay area contains the engine control room, instrument panels, high-voltage transformers, and power-distribution equipment.

The system energy flow diagram is shown in Figure 8. The continuous electrical output rating of the system is 14,000 kW (net). Residual-fuel-oil consumption is ap-

TABLE 5

Cogeneration System Performance Summary

Net electric output	14,000 kW
Residual oil consumption	6818 lb/hr
	122.7×10^6 Btu/hr (LHV)
Specific fuel flow rate	0.487 lb/kWh
Overall plant electrical efficiency	38.9%
Engine exhaust conditions	
Specific flow rate	22.5 lb/kWh
Specific heat	0.254 Btu/lb °F
Exhaust-gas temperature	626°F
Exhaust heat boiler	
Gas inlet temperature	608°F
Gas outlet temperature	350°F
Specific steam flow rate	1.42 lb/kWh
Steam pressure	90 psig
Specific heat recovery	1474 Btu/kWh
Engine cooling heat exchanger	
Water outlet temperature	180°F
Specific heat recovery	1936 Btu/kWh
Energy balance	
Fuel input	8771 Btu/kWh
Gross electricity (96% gen. eff.)	3483 Btu/kWh
Net electricity (2% auxiliary load)	3413 Btu/kWh
Steam recovery	1474 Btu/kWh
Hot water recovery	1936 Btu/kWh
Total energy utilization	6823 Btu/kWh
Stack heat loss	1355 Btu/kWh
Engine radiation loss	172 Btu/kWh
Exhaust manifold loss	103 Btu/kWh
Generator cooling loss	145 Btu/kWh
Water heat exchanger loss	103 Btu/kWh
Auxiliary load	70 Btu/kWh
Total losses plus auxiliary load	1948 Btu/kWh
Energy utilization (%)	77.8
Cogeneration heat rate	5361 Btu/kWh

proximately 19.5 b/hr (122.7×10^6 Btu/hr). The energy content of the steam and hot water produced from the reject heat of the diesel totals to 47.7×10^6 Btu/hr. If this process heat were produced in a separate boiler, an additional 8.9 b/hr of fuel oil would have to be consumed. A summary of the system performance data is presented in Table 5. The total energy utilization is 6823 Btu/kWh, or 77.8% of the energy input. In terms of electrical generation, the effective cogeneration heat rate is 5361 Btu/kWh.

VII. SYSTEM ECONOMICS

Estimated costs for the installation and operation of the proposed diesel cogeneration system are described below. The costs are in terms of 1977 dollars and represent a fully installed operating plant. Equipment costs include:

Plant section	Cost (1977 $)
Diesel engine	253/kW
Engine room accessories	46/kW
Generator	89/kW
Engine shipping	24/kW
Engine erection and commissioning	30/kW
Subtotal	442/kW
Diesel-generator (14,000 kW)	6.2×10^6
Exhaust recovery boiler and hot water heat exchanger	0.3×10^6
Steam, water piping, and exhaust ducting	0.1×10^6
Electrical switchgear, transformers, controls, and cabling	0.3×10^6
Building and stack	0.6×10^6
Modifications to electrical distribution	0.5×10^6
Installed equipment costs	$8.0 \ \times 10^6$
Engineering and design	$1.0 \ \times 10^6$
Installed system costs	$9.0 \ \times 10^6$
General and administrative expense (7%)	0.63×10^6
Total installed system costs	9.63×10^6

Operating costs for the complete system are summarized below. They are annual costs, assuming a 0.9 utilization factor and energy costs of $1.91/$10^6$ Btu (LHV) for residual oil.

Variable	Cost (1977 $)
Residual oil ($0.9 \times 8760 \times 122.7 \times 10^6 \times 1.91 \times 10^{-6}$)	1.85×10^6
Property taxes and insurance ($2\% \times \$9 \times 10^6$)	0.18×10^6
Maintenance and expendables ($2\% \times \$9 \times 10^6$)	0.18×10^6
Burdened labor ($5 \times \$20,000$)	0.10×10^6
Total annual operating costs	2.31×10^6

Annual pretax savings derived from the system operation are calculated as the difference between the operating costs and the value of equivalent amounts of purchased electricity and process heat produced in existing boilers. A value of $0.03/kWh is taken for purchased electricity, and the cost of fuel adjusted for boiler efficiency is used for process heat. Accordingly, the annual savings are as follows:

Variable	Cost (1977 $)
Value of electricity ($0.9 \times 8760 \times 14,000 \times 0.03$)	3.31×10^6
Value of process heat ($0.9 \times 8760 \times 46.1 \times 10^6 \times {}^{1.91}/_{0.85} \times 10^{-6}$)	0.82×10^6
Total value of electricity and process heat	4.13×10^6
Annual pretax savings ($4.13 \times 10^6 - 2.31 \times 10^6$)	1.82×10^6

The internal rate of return of the system investment can be calculated using these pretax annual savings at time = 0, assuming all costs and values escalate 6%/year (continuous compounding) and assuming the following:

Federal/state tax rate	50%
Investment tax credit	10%
Plant life	20 years
Sum-of-digits accelerated depreciation allowed	15 years
Salvage value	0

The resultant effective annual rate of return based on a continuous cash flow model is 21.0%.

VIII. SUMMARY

The diesel engine is the highest efficiency prime mover commercially available. Basic engine efficiencies range up to as high as 42% relative to shaft output. Net efficiency after accessories and generator losses are in the 38% range for engines larger than ∿3000 hp.

The ratio of power output to process steam is ∿400 to 700 kWh/10^6 Btu, or about 10 times greater than that attainable utilizing a back-pressure turbine with typical system conditions. The diesel is therefore particularly well suited to cogeneration situations requiring a relatively high proportion of electric power compared to thermal products. In typical cogeneration applications, a fully installed diesel system is roughly in the cost range of 500 to 800 $/kWe (1977 dollars). Overall fuel utilization efficiency for power and thermal products can exceed 75%.

Medium- and slow-speed diesels are capable of running on continuous diets of residual oil. Fuel treatment is normally required if vanadium content exceeds 50 to 100 ppm and/or sodium content exceeds either one third the vanadium content or 15 ppm. Programs are currently underway to develop diesel engines which will utilize coal-derived fuels, coal-pil slurries, and hopefully, pulverized coal directly.

REFERENCES

1. **Bryant, Lynwood,** Rudolf Diesel and his rational engine, *Sci. Am.,* 221, No. 2, 1969.
2. **Anon.,** High Efficiency Decentralized Electrical Power Generation Utilizing Diesel Engines Coupled with Organic Working Fluid Rankine Cycle Engines Operating on Diesel Reject Heat, TE4186-27-75, Thermo Electron Corporation, Waltham, Mass., for the National Science Foundation, Washington, D.C., November 1974.
3. **Carnot, S.,** *Reflections on the Motive Power of Fire,* Dover Publications, New York, N.Y., 1960.
4. **Diesel, R.,** U.S. Patent 542846, 1895.
5. **Pawlikowski, R.,** The coal dust engine upsets traditions, *Power,* 136, 1928.
6. **Morrison, L. H.,** The coal dust engine, *Power,* 746, 1928.
7. **Schreck, H.,** Burning powdered coal in Diesel engines, *Power,* 748, 1928.
8. **Anon.,** The Rupa pulverized fuel engine, *Engineering,* 408, 1928.
9. **Anon.,** Powdered fuel engine tests, *Engineer,* 166 (4328), 700, 1938.
10. **Anon.,** Pulverized Coal Burning Diesel Engine, Research and Development Rep. No. 46, Contract No. 14-01-0001-491, Office of Coal Research, U.S. Department of the Interior.
11. Final Report — A Study of Inplant Electric Power Generation in the Chemical, Petroleum Refining and Paper and Pulp Industries, TE5429-97-76, Thermo Electron Corporation, Waltham, Mass., for the Federal Energy Administration, Washington, D.C., 1976.

Chapter 9

FUTURE TECHNOLOGIES

John P. Ackerman and Robert E. Holtz

TABLE OF CONTENTS

I. INTRODUCTION

The increasing price and decreasing availability of oil and natural gas provide the incentive for development of technology to use these premium fuels more effectively, and to replace them with other energy sources. Coal and uranium are more plentiful than oil and gas. Solar and geothermal heat, as well as agricultural, industrial, and urban wastes, are all renewable.

In order to conserve and replace oil and gas through cogeneration, it is desirable to develop technologies which can employ a host of fuel options and which have other characteristics which increase the number of attractive cogeneration situations. Among these characteristics are the capability of operation in previously impractical temperature ranges, flexibility of output, low cost, etc. Among the advanced technologies of interest are several types of fuel cells, coal-burning diesel engines, high-temperature directly fired gas turbines, externally fired gas turbines, Stirling engines, and organic Rankine-cycle engines. This chapter is intended to provide an introduction to these technologies, with particular emphasis on fuel cells as a sample case.

II. FUEL CELLS

A. Description

Fuel cells are a unique class of devices for direct conversion of chemical energy into electricity. They have none of the intermediate conversion steps (chemical energy to heat, heat to mechanical energy, and mechanical energy to electricity) that are characteristic of combustion engines. In any fuel cell, a fuel gas is oxidized electrochemically by releasing electrons to the cell anode (see Figure 1), and an oxidant (nearly always air) is reduced by accepting electrons from the cathode. The net result is the flow of electrons from anode to cathode (generation of electric power) and oxidation of the fuel. A flow of charged ions in the electrolyte completes the electrical circuit. Fuel cells are classified by the nature of the ionic species in the electrolytes. Some of the more important types are discussed below.

Cell output is typically from 0.5 to 1.0 volt DC, so on the order of 1000 individual cells are connected in electrical series to form a "stack" of cells with a useful voltage output. The electrical conductor between cells serves to separate the fuel gas of one cell from the oxidant of the next. Therefore, external wiring is minimized, and the stack becomes a compact, modular unit. All the cells conduct the same current, which can vary instantly from zero to several hundred mA/cm^2 of cell area in response to load demand. Practical cell areas are typically 0.1 to 1.0 m^2. Where AC power is required, inverters of about 97% efficiency are presently available.

One major advantage of fuel cell generators is that they produce essentially no pollutants. Since there is no combustion in the cell, no NO_x or soot are formed. All practical fuel cells (or their fuel processing equipment) are sulfur sensitive, so sulfur is typically reduced to very low levels in the fuel stream. Depending on the fuel source, some combustion may be practiced to provide heat for fuel processing. Usually the depleted fuel stream from the cell anode exhaust is used for this purpose. Therefore, combustion temperatures are low, and NO_x formation is negligible.

The cells themselves have no moving parts, so the noise generated is that associated with pumps and blowers or with fuel processing equipment.

Unlike other prime movers, the ultimate efficiency of fuel cells is not limited by the Carnot relationship for heat engines where

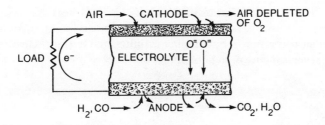

CELL REACTIONS

CATHODE $\quad 2e^- + \frac{1}{2} O_2 \rightarrow O^=$

ANODE $\quad\quad CO + O^= \rightarrow CO_2 + 2e^-$

$\quad\quad\quad\quad OR \; H_2 + O^= \rightarrow H_2O + 2e^-$

OVERALL $\quad CO + \frac{1}{2} O_2 \rightarrow CO_2$

$\quad\quad\quad\quad OR \; H_2 + \frac{1}{2} O_2 \rightarrow H_2O$

FIGURE 1.   Section of a fuel cell (oxide electrolyte type).

$$\eta = \frac{T_u - T_1}{T_u} = \frac{\text{mechanical work}}{\text{heat input}}$$

The symbol η is the maximum possible conversion efficiency of the heat engine, T_u is the temperature at which heat enters the device, and T_1 at which heat is rejected. T_1 is constrained to be above ambient temperature for heat rejection. Heat engines can be used as either "topping" or "bottoming" devices for cogeneration. When they are used as topping devices, T_1 must be selected to meet the industrial process requirement. In the bottoming mode, the industrial process decreases the temperature of a fluid stream, and the heat engine produces work by employing some of the remaining stream heat. In this case, T_u is the temperature at which heat leaves the industrial process.

Fuel cells operate in a fundamentally different manner. They are not heat engines. The Carnot relation does not apply. The source of their energy is not heat, but rather, the chemical energy of oxidation of a fuel. Like other conversion devices, they change part of the fuel chemical energy into electricity, but also produce some heat. The critical difference from heat engines is that this heat is released at the cell operating temperature. The temperature (but not the quantity) of the heat released is substantially independent of the amount of electricity generated.

The heat and electrical output of a fuel cell can be calculated from a knowledge of the cell operating voltage, V, and the thermodynamic properties of the reactant gases. The electrical energy output of a cell which consumes 1 mol of fuel is VnF, where V is the cell voltage, n is the number of equivalents per mol of fuel, and F is Faraday's constant, 96,500 C/equivalent. The energy content of that mol of fuel is conventionally taken to be ΔH, the enthalpy change in the cell reaction. The maximum electrical energy per unit heat output is obtained when the cell operates under infinitesimal load (current). Under these conditions, the cell voltage is at a maximum value V_E. V_E is related to the Gibbs Free Energy change, ΔG, of the mol of fuel in the cell reaction by

$$\Delta G = V_E nF$$

The ratio $\Delta G/\Delta H$ is like the Carnot limiting efficiency of heat engines. The values of $\Delta G/\Delta H$ range from about 70% to over 100%, i.e., some cells can theoretically abstract heat from their surroundings if operated at infinitesimal current. Under real loads, the actual ratio of electrical energy to heat output is*

$$\frac{VnF}{\Delta H - VnF}$$

The mol of fuel referred to above is nearly always accompanied into the cell by some extra fuel which passes through the cell unchanged. The reason for this is that the cell voltage is degraded at very low fuel concentrations. These can occur near the exit port of the cell if the fuel is diluted by inerts or, more significantly, by reaction products. A similar situation obtains at the cathode, but the fraction of oxidant flow converted by the cell is usually relatively low because air is cheap. Also, the cathode flow is sometimes used for heat transfer out of the cell. The electrical energy/heat ratio, assuming combustion of unconsumed fuel, is, therefore,

$$\frac{VnF \, X}{\Delta H - VnFX}$$

where X is the fraction of fuel consumed.

Polarization curves are plots which relate V to the current (or power, which is the product of the current and voltage) produced by the cell. The form of the polarization curve is determined by the kinetics of the cell processes. The reader is referred to several excellent texts for a full discussion.[1,2,3] Figure 2[4] shows a family of polarization curves for a fuel cell of the molten carbonate electrolyte type. The anode feed is a mixture representative of coal gasifier output containing inerts, CO_2 and H_2O (products), and H_2 and CO (fuel). Operating pressure is 1 at, although cell voltage would be about 0.1 V higher had the cell been operated at 5 to 10 at.

The horizontal axis is plotted as "current density" or current/unit cell area, because cell performance is essentially independent of cell size. A cell with a few cm² area will have the same voltage to current density relationship as a practical cell whose area may be 10^3 or more times as great. This property contrasts sharply with the behavior of conventional heat engines, such as steam and gas turbines, etc., where the efficiency of small engines is nearly always less than that of large units. The size of the fuel cell stack can, therefore, be chosen to maximize convenience of shipping and/or installation, thereby reducing installation cost and lead time substantially. Modularity also allows the fuel cell power plant to be readily sized for each individual application and to be expanded as the need arises.

Fuel cells provide the power plant designer with considerable flexibility to vary the electrical/heat output ratios independently of total thermal plus electrical demand. The maximum-design electrical-to-thermal ratio depends on the total cell area used to meet a given demand (high cell area means low current density and, therefore, high relative electrical output). It is also possible to vary the electrical output independently of fuel input. Finally, the fuel input rate can be variable, depending on the fuel processor, to cope with changing total energy requirements throughout the duty cycle of the power plant. Generally, the response of the fuel cell to load change is very rapid (< 0.1 sec). Therefore, response-time will be determined by the fuel processor. The potential of certain fuel cell types for industrial cogeneration is just now being realized, and conceptual designs are underway.

---

*Note that as the electrical load approaches zero, this expression approaches ΔG, the cell efficiency.

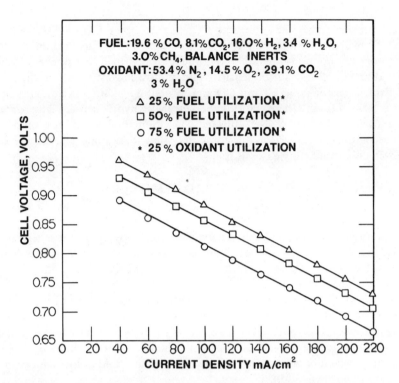

FIGURE 2.   Molten carbonate fuel cell performance, simulated coal gasifier product.

Several types of fuel cells are of interest for industrial cogeneration. Cells with phosphoric acid electrolyte are on the threshold of commercial use for electric utility power and for residential total energy systems. A 1 MW generator has been operated successfully, and a 4.8 MW demonstration plant is slated for operation in New York City within the year. Cell voltages are in the 0.5 to 0.75 V range. Unfortunately, cell temperature is only about 450 K, and this limits industrial cogeneration applications to those requiring only relatively low temperatures.

Cells with an electrolyte composed of molten alkali carbonates are about 5 to 7 years behind phosphoric acid cells in development. They will operate in the 0.5 to 1.0 V range of cell voltage, with a cell temperature near 925 K. They are well-suited to operation on a variety of fuel sources, including coal gasifiers, and may be used in a great many industrial cogeneration applications.

Cells with solid oxide electrolytes operate at temperatures near 1275 K on a variety of fuels, but are considerably less well-developed than the other two. Potentially, they are the most interesting of all due to their simplicity and high temperature.

B. Fuel Requirements

Fuel cells operate by oxidation of a fuel gas, such as H_2 or CO. Any material, such as coal, which can be converted into H_2 or CO is potential fuel. Each type of cell has specific requirements which are discussed here.

Phosphoric acid cells operate on H_2 only, and will not accept CO levels above ∼1%. H_2 is typically obtained by the reforming of carbonaceous fuel, followed by shift reaction of CO and H_2O to produce H_2 and decrease CO concentration. The source of H_2 could equally well be biomass, coal, industrial waste streams, etc. H_2S at the 100 ppm level is tolerable by the cell, but shift reaction catalysts are intolerant of ppm sulfur levels.

Molten carbonate fuel cells consume CO, in fact, some CO or CO_2 in the fuel is required. They are well-suited for operation on coal gas or any carbon-bearing fuel. Hydrocarbons are not consumed directly, but reformed fuel does not require shifting. Molten carbonate cells are sulfur-sensitive and require sulfur removal to the ppm level.

Solid oxide cells consume H_2 and CO and, probably, significant amounts of hydrocarbons as well, although there is little experience with hydrocarbons. The contaminant tolerance also has not been well-established. Developmental emphasis is on other areas at the present stage. It is reasonable to assume that these high temperature cells will operate well on virtually any gaseous fuel.

III. OTHER PRIME MOVERS

A. Stirling Engines[5,6,7]

Large, stationary Stirling engines are external combustion machines that offer several advantages over conventional engines. These include greater fuel efficiency, greatly reduced environmental impacts (noise and emissions), and the capability of using alternate, nonscarce fuels.

The ideal Stirling engine operates on a thermodynamic cycle having a theoretical efficiency equal to that of the Carnot cycle. The Stirling cycle is composed of the following processes: (1) isothermal expansion in which heat is transferred to the working fluid at high temperature, (2) constant-volume heat extraction, (3) isothermal compression in which heat is removed from the working fluid at low temperature, and (4) constant-volume heat addition.

Practical Stirling engine designs utilize regenerators and various piston arrangements to perform the four processes. Thermal efficiencies in excess of 40% should be obtainable in advanced Stirling engines.

A major asset of the Stirling engine is that it can be employed with a variety of heat sources other than high-grade petroleum or natural gas fuels. Among the energy sources under consideration are low-grade petroleum residues, coal-derived liquids and gases, fluidized-bed combustors, and municipal waste incinerators. In addition to their multifuel capabilities, Stirling engines have better efficiencies (particularly at part-load conditions) than conventional systems.

A major factor influencing the use of Stirling engines for industrial cogeneration systems is the uncertainty of future scarce fuel resources. The development of a large Stirling engine with its inherent fuels flexibility would encourage industrial cogeneration applications. Using the Stirling engine either as a topping or a bottoming cycle in an industrial cogeneration system can result in overall system efficiencies of around 80%. Figure 3 illustrates an industrial cogeneration system where the Stirling engine is used for topping, and usable heat is available for industrial processes.

Most of the Research and Development effort on Stirling engines has been concerned with units of less than 300 hp in size. These applications have ranged from small units for power generation or use with heat pumps to Stirling engines for vehicular transportation. The primary requirements for these engines are considerably different than the larger size engines that appear to be needed for industrial cogeneration applications. Efforts to develop large, stationary Stirling engines suitable for industrial cogeneration applications have been initiated only recently. These new developmental efforts are directed to Stirling engines in the 500 to 2000 hp range. Demonstration of Stirling engines in this size range is anticipated to take place in 1985.

B. Organic Rankine Cycle[8]

Organic Rankine cycle engines are being developed for potential use in industrial

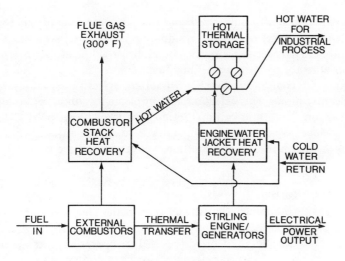

FIGURE 3.   Stirling engine-based industrial cogeneration system.

cogeneration systems. These engines will be sized under 1 MW. These engines are ideal for recovering waste heat from diesel engines and other prime movers, and for other waste-heat recovery applications in industry. These engines can recover the waste heat from prime mover exhausts and convert it to additional useful shaft power at efficiency levels of 18 to 20%.

Organic Rankine cycle engines are thermally driven devices that convert heat energy into mechanical energy by alternately evaporating a working fluid at high pressure and producing shaft power from the high pressure vapor as it expands through a turbine to a condenser which operates at low pressure. Simply, the waste-heat recovery system consists of a vaporizer, a control system, and a power conversion system.

Vaporizers are tailored for each application. They include a diverter valve capable of diverting the heat source flow to a bypass system. The control system consists of all controls, except those mounted on system components. The power conversion system includes the remaining system components, including turbine and feed pump, regenerator, condenser, gear box, noncondensible removal system, hotwell, booster pump, starting pump, flow control valve, generator, and miscellaneous wiring, piping, valves and fittings.

These engines, being developed by Mechanical Technology, Inc., Sundstrand Energy Systems, and Thermo Electron Corporation, are in the range of 400 to 600 kW and will be field tested prior to 1980. The working fluids in these systems are Freon®, toluene, and Fluorial® 85.

C. Coal-Burning Diesel[9,10,11,12]

Present-day large-bore, slow-speed diesels, with characteristics of high-efficiency, minimum cooling-water requirements, competitive costs, short construction time, and potential for operation on coal and coal-based fuel, appear very attractive for fulfilling power requirements for industrial cogeneration. For cogeneration, the diesel offers approximately twice the electricity output per unit of steam output as the gas turbine and 10 times that of the steam turbine.

The current diesel technological development activities include the utilization of alternate fuels, including petroleum residual and coal-based fuels, for cogeneration. This fuels-oriented effort is structured to provide for a progressive implementation of engine modifications and redesigns aimed at achieving engine verification on residual fuels in the near-term, and coal-based fuels in the long-term.

Early German experience with pulverized coal and more recent experience with slow speed, large-bore marine diesel engines on residual fuels indicate that today's slow-to-medium speed stationary diesels have much potential for achieving near-term acceptance in the industrial sector by burning residual fuel.

Keys to achieving the coal-based diesel fuel goal are effective use of materials (ceramic) and lubricants, along with selection and implementation of optimum engine (injection, combustion, and lubrication) design features. Slow-speed diesels with larger combustion chamber space offer opportunities to reduce nitrogen oxides by applying radically new fuel injection and ignition sources to achieve proper control of the combustion process.

D. Externally Fired Gas Turbines[13,14,15,16]

Externally fired gas turbines have the advantage of being able to use alternate, non-scarce fuels, including solids, liquids, and gases. The open, externally fired Brayton cycle has the additional advantage of providing high temperature pure air for use in cogeneration applications. The closed externally fired Brayton cycle has the alternate advantages of low noise and small equipment size because of a high-pressure working fluid.

The energy conversion efficiency for Brayton cycles depends on the cycle pressure ratio and cycle temperature. The thermodynamic cycle of an open externally fired Brayton cycle includes the following processes: (1) isentropic compression, (2) heat addition at constant pressure via heat exchangers and an external combustion chamber, and (3) isentropic expansion through the turbine blades. The gases are exhausted to the environment after expansion through the turbine unless used for additional heat recovery. The thermodynamic cycle of closed externally fired Brayton includes the processing of the open cycle with the addition of a fourth process, heat rejection at constant pressure, to complete the cycle ready for the isentropic compression.

Thermal efficiencies for existing externally fired gas turbines are comparable to those of direct-fired systems with equivalent pressure ratios and turbine temperatures. These have upper values in excess of 30%. Today's commercially available units of less than 20 MW generally have efficiencies of less than 25%. The principal limitation to increased efficiency in externally fired gas turbines is the low turbine-inlet temperatures that can be achieved because of metal temperature limitation in current heat exchangers.

The externally fired gas turbine can be used with a variety of heat sources other than high-grade petroleum or natural gas fuels. Among the energy sources under consideration are low-grade petroleum residues, coal-derived products, fluidized-bed combustors, and waste heat from various sources. The goal of improved efficiency will take place along two fronts. The first is improved efficiency of the prime mover through development of heat exchanger technology for higher turbine inlet temperatures and improved cycle development. The second front includes the use of the externally fired gas turbine in cogeneration or combined cycles. Cogeneration systems of this type can result in overall system efficiencies as high as 70 to 80%.

In addition to the externally fired units which range up to 20 MW, the development of coal-fired primary heater designs for closed-cycle gas turbines in the 25 to 50 MW range are underway. This effort addresses both metallic-surface primary heaters for turbine inlet temperatures ranging up to 1550°F and primary heaters with ceramic surfaces for inlet temperatures ranging up to around 1750°F.

E. Directly Fired Gas Turbines

Open cycle, directly fired gas turbines have historically been reliable, high power-density prime movers with low acquisition costs. However, existing gas turbines rely on critical gas and distillate fuels that may become increasingly scarce past the mid-1980s. This scarceness of critical fuels could force discontinuing the use of gas turbines for some applications unless alternate fuels can be employed.

The use of residuals as turbine fuel would offer an effective means of conservation by lessening the competition for the lighter distillates which are needed for home heating and in the transportation sector. The problems of burning residual oil in gas turbines are similar to those anticipated with future coal-derived liquid fuels. If developed in an appropriate manner, the resulting technology would permit heavy-fuel turbines in the field ultimately to be converted to burn coal or oil-shale-derived fuels when they become available in sufficient quantities.

Some of the development efforts focus on technological aspects of improved performance and fuel flexibility. These include thermal barrier coatings for turbine protection; heavy fuel, low nitrogen oxide combustor technology for burning fuels high in fuel-bound nitrogen while meeting nitrogen oxide emission standards; advanced convectively cooled turbine airfoils for enhanced cooling and improved durability with ash-bearing low-grade fuels; and advanced low-aspect-ratio compressor technology to maximize cycle efficiency through increased pressure ratios and to minimize the number of parts and complexity relative to existing compressors.

IV. SUMMARY

The salient characteristics of some of the most promising advanced technologies have been discussed. The capability to use a variety of fuels and/or energy sources is a property common to all. It is achieved in several ways. The Stirling engine, organic Rankine engine, and externally "fired" gas turbine require only some source of external heat. The coal-burning diesel and directly fired gas turbine are being developed to use low-grade fuels through materials and design improvements. Fuel cells can use a wide variety of fuels because of external fuels processing. Fuels flexibility is clearly a major factor in the struggle to reduce dependence on the dwindling supply of premium fuels. Each of these technologies has some other special characteristics which can be used to extend the range of applications of industrial cogeneration systems, and thereby, decrease the energy resources required to support the industrial sector.

REFERENCES

1. Newman, J., *Electrochemical Systems,* Prentice-Hall, Englewood Cliffs, N.J., 1973.
2. Liebhafsky, H. and Cairns, E., *Fuel Cells, a Guide to Their Research and Development,* John Wiley & Sons, New York, 1968.
3. Bockris, J. and Srinivasan, S., *Fuel Cells; their Electrochemistry,* McGraw-Hill, New York, 1969.
4. Anon., Institute of Gas Technology Project 8984 Final Status Report, for the Argonne National Laboratory, Argonne, Ill., September 30, 1977.
5. Martini, W. R., Stirling Engine Design Manual, NASA CR-135382, National Aeronautics and Space Administration, Washington, D.C., April, 1978.
6. Walker, G., *Stirling Engines,* Vol. 1 and 2, Oxford University Press, Oxford, England, 1979.
7. Walker, G., *Stirling Cycle Machines,* Clarendon Press, Oxford, 1973.
8. Davis, J. P., Reciprocating engines with heat recovery, in *Cogeneration of Electricity and Useful Heat,* CRC Press, Boca Raton, Fla., 1980.
9. Anon., Assessment of Technology for Advanced Power Cycles, NAS Rep., National Academy of Science, Washington, D.C., 1977.

10. **Soehngen, E. E.**, Development of Coal-Burning Diesel Engines in Germany: a State-of-the-Art Review, Soehngen & Associates, Fairborn, Ohio, 1976.

11. **Tipler, W.**, *Prospects for the Operation of Diesel Engines on Coal or its Derivations, Power Plants and Future Fuels,* Mechanical Engineering Publications, Ltd., New York, 1975.

12. **Raymond, R. J., Morgan, D. T., Appleton, J. P., and Davis, J. P.**, Cost and Application of Coal-Burning Diesel Power Plants, TE2557-110, Thermal Electron Corporation, Waltham, Mass., 1975.

13. **Schweitzer, J. K. and Brown, B. T.**, Advanced Industrial Gas Turbine Technology Readiness Demonstration Program, HCP/T5035-0001, Pratt & Whitney Aircraft Group, for the U.S. Department of Energy, 1977.

14. **Farahan, E. and Eudaly, J. P.**, Gas Turbines, ANL/CES/TE 78-8, Argonne National Laboratory, Argonne, Illinois, 1978.

15. **Smith, I. E.**, *Combustion in Advanced Gas Turbine Systems,* Pergamon Press, Oxford, 1967.

16. **Sorensen, H. A.**, *Gas Turbines,* The Ronald Press Co., New York, 1951.

Project Implementation Considerations

SECTION 3

PROJECT IMPLEMENTATION CONSIDERATIONS

PREFACE

The three chapters in this section are intended for those readers who may be actively considering the design and/or installation of a cogeneration project. Identified herein are many (if not most) of the factors which must be evaluated and harmonized to implement various types of cogeneration systems.

Chapter 10 discusses central utility plant cogeneration operations primarily with respect to district heating systems. New plant installations as well as the retrofit of existing plants are covered. The step-wise approach to new projects used in some European areas is described. The major barriers to implementation are presented, and the chapter concludes with some basic economic considerations.

Chapter 11 presents a review of the significant installation requirements of both large and small industrial systems. In the case of large systems, special emphasis is given to the technical requirements of coal-fired systems and to the interconnection with a local utility. The small-plant discussion deals largely with the techniques used to minimize costs and ensure reliability in a small installation.

Chapter 12 identifies the major options associated with a community Total Energy system, together with the factors which influence their selection. Consideration is given to community characteristics, energy load characteristics, technology alternatives, and system performance. The chapter also presents some observations on coping with the legal and regulatory conditions affecting a Total Energy project.

Chapter 10

TECHNICAL AND ECONOMIC PARAMETERS FOR CENTRAL UTILITY BASED SYSTEMS

Thomas E. Root

TABLE OF CONTENTS

I. INTRODUCTION

For many years, the American people were accustomed to cheap and reliable sources of energy to satisfy all of their needs and desires. It was not until recently that the limitations on fossil fuel reserves were brought home to the general public. This realization has brought about an intensive search for alternate energy sources and more efficient means of utilizing fossil fuels. District heating, via cogeneration, is one of the recognized methods for conservation of energy resources. It is not, however, a panacea. Although district heating uses existing technology, it is neither fast nor easy to install. Most electric generating plants cannot easily be retrofitted as a heat source. Many existing buildings cannot be easily retrofitted to accept district heating. How, then, can a cogeneration-based district heating system be installed economically with a reliable supply of heat to the customer? When and where is it applicable?

II. ENERGY CONSERVATION

The potential energy savings which can be realized through district heating via cogeneration are in the range of 30 to 35%, based on European experience. However, care must be taken in applying these savings to potential systems. These savings cannot be applied to the total U.S., or even to the energy consumption for space and water heating in a single country. Only areas with sufficiently high load densities and an existing power or steam plant which can be efficiently converted to cogeneration, or which can justify construction of a new cogeneration plant, can realize such large savings and justify a district heating system.

Many of the best locations for district heating systems are in the urban areas where heat-load densities are very high. Unfortunately, most existing power plants in these areas are among the oldest plants on the electrical system and burn oil or natural gas. Both of these conditions limit the potential energy savings realizable. In some cases, retrofitting of these plants for district heating may not be economically justifiable because of the age of the equipment or the design of the plant. The substitution of district heating based on oil and natural gas for the in-building systems, based on the same fuels, does not result in the most economical or productive use for these fuels.

III. FUEL CONSIDERATIONS

The choice of fuel or fuels for the cogeneration plant impacts its economic viability and environmental acceptability. Locating the plant in or near the service area, which is the ideal choice from an economic point of view, weights heavily in favor of natural gas and oil as the primary fossil fuel candidates. Coal is greatly restricted as a viable option, and nuclear fuel is impossible under current conditions. However, two additional, nonconventional fuels, refuse and wood, offer possible alternatives.

Some of the advantages and disadvantages of each type of fuel are shown in Table 1. While oil and natural gas are both environmentally acceptable and easily stored, they suffer from high cost and the potential of curtailment of supply. Coal is the least expensive and most readily available fuel, but had severe environmental and public acceptance limitations. However, if an existing coal-fired plant can be retrofitted or rebuilt for congeneration, then it would be a prime choice for becoming a base-load plant. Municipal refuse presents a unique opportunity in that it is generated in the service areas and is renewable. However, because of its low density it requires relatively large boilers and may present problems in maintaining steam conditions in the turbine. Wood, like refuse, can be obtained from the local area and is clean burning. Even

TABLE 1

Fuel Choices

Fuel	Advantages	Disadvantages
Gas	• Environmentally acceptable • Easily transportable • No storage required • Small plant size	• Expensive • Availability
Oil	• Environmentally acceptable • Relatively easy transport • Easy storage • Small plant size	• Expensive • Availability
Coal	• Inexpensive • Availability	• Transportation by rail or water to urban area • Precipitators will, and scrubbers may, be required • Large storage area required • Large plant size • Public acceptance in urban area
Refuse	• Relatively inexpensive • Environmentally acceptable • "Renewable" supply • Reduced land-fill requirements	• Large plant size • Inconsistencies in heating value • Limited supply • Heavy truck traffic • Low density
Wood	• Relatively inexpensive • Environmentally acceptable • Semi"renewable" resource	• Collection and transportation • Storage and handling • Limited supply • Limited applicability

large metropolitan areas generate substantial quantities of waste wood from tree trimming, scrap pallets, etc. The greatest limitations for both refuse and wood are their relatively limited supply.

If the plant is located in low population areas some distance from the service area, then both coal and nuclear plants can be seriously considered. However, these plants tend to be the largest plants in the electrical system and, thus, are only practical for cogeneration if the district heating system is large, or if it is not feasible to locate the cogeneration plant near the service area.

IV. SYSTEM DEVELOPMENT AND EXPANSION

The utility-based district heating system is designed to serve a loosely defined area with changing needs, including expansion. This differs from other systems which are designed to serve a specific, well-defined load and area, such as a hospital or shopping mall. Because of these differences, the district heating system is structured very much like an electric utility.

The system can be divided into base-, intermediate-, and peak-load plants. The plant mix is determined primarily by the seasonal load characteristics for the system and, therefore, must be established on a case by case basis. Nominally, a system would consist of 50% base-load plant, 30% intermediate-load plant, and 20% peaking.[1] Cogeneration plants are most applicable to base-load use. This assures high utilization of capital-intensive equipment. The intermediate-load plants should be stationary boilers, including refuse plants. Here, economy of scale can be realized without the additional capital cost of a turbine generator on a plant that will not have a high load factor. Peaking requirements may best be satisfied by portable boilers. These highly versatile units are discussed in more detail below.

In an expanding district heating system, new, large blocks of load do not materialize suddenly. They grow gradually as an area expands and develops, or is rebuilt. Unless the existing district heating system is very large and has units which should be retired, the construction of a cogeneration plant to serve the new area cannot be economically justified in the early stages of area development. However, if district heating is to serve the area, it is most economic to install the district heating lines and building heat exchangers at the same time as the other utilities are installed, i.e., during building construction. The Europeans have developed a three-step system which provides a smooth and economic transition from a newly developing area with low load to incorporation into the district heating system. The steps are

1.   Remotely controlled portable boilers are installed near the largest users in the developing area. Since the larger users are generally the first to be connected, they serve as the central islands to which other buildings in the area can be connected as they are completed.
2.   Several of the islands are interconnected, and a permanent heat-only boiler plant is constructed to serve the area. Most of the portable boilers are then removed to new areas of development. However, some may be retained to serve as peaking and back-up for the boiler plant. At the same time, mains are run to interconnect the main district heating system with the newly developed area.
3.   Finally, a new cogeneration plant is built to maintain the cogeneration capacity at approximately 50% of the peak load. The heat-only boilers, which have lower capital costs, but higher fuel costs, are then used for intermediate and peak loads and as reserve units for the cogeneration plant.

The timing of the second and third steps, and total user participation in the system, are very important if the overall network costs are to be kept low. Because of regional variations in construction and operating costs, the lowest acceptable capacity factor on the units must be analyzed on a case by case basis. The capacity factor will, of course, rise as the system grows.

The flexibility of the portable boilers is worthy of some additional comments. Because of their small size and high portability (the European hot water boilers resemble a small mobile home with a guy-wired stack), they can be quickly and easily be moved to new locations in emergency situations. This portability also makes shop maintenance feasible.

V. CONVERSION OF EXISTING SYSTEMS TO COGENERATION AND DISTRICT HEATING

There are two distinct aspects of existing heating systems which must be considered in conversion to district heating through cogeneration. One is the ability to connect the heating systems currently employed by the user. The other is the ability to convert existing power plants to cogeneration as a source of heat for the district heating system.

The conversion of most existing heating systems to either a steam or hot water district heating system is, technically, relatively easy. If the existing system is forced air, then a water to air, or steam to air, heat exchanger can be installed to replace the existing furnace. Likewise, a hot water boiler can be replaced with a steam to water, or water to water, heat exchanger. However, it is not possible to easily replace an

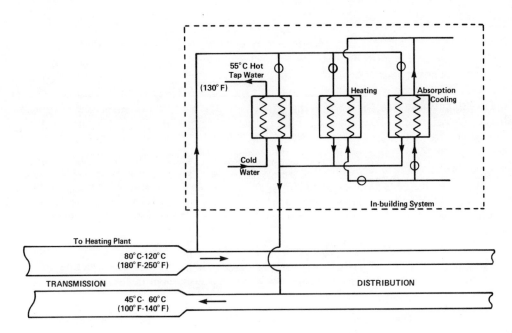

FIGURE 1.   User connection for hot water heating and cooling system.

existing steam heating system with a hot water district heating system because the radiators and some piping and valves would need to be replaced in addition to the boiler. One system which does not lend itself to conversion to district heating is baseboard electric resistance heating. Because there is no central heat source and, thus, no duct work or piping, it is impractical to retrofit buildings equipped with baseboard electric heat to a district heating system. Heat pumps can be indirectly connected to a district heating system by installing heat exchangers on the outside condenser unit to preheat the air. This raises the coefficient of performance.

Conversion of existing buildings to district heating and cooling will require modification of three systems by the building owner: heating, cooling, and hot tap water. The degree of difficulty of the conversion is dependent upon the design of the existing systems and of the building.

A typical arrangement of user equipment is shown in Figure 1, based on a year-round supply of hot water from the heating plant. Two heat exchangers are used to supply hot tap water and heating. These replace the existing boiler or furnace and hot water supply. In Europe, these two systems are incorporated into one tank, but they have no air conditioning load because of a moderate summer. For U.S. applications, it will probably be more efficient to separate these functions if a significant air conditioning load is to be served. An additional heat exchanger, utilizing the return water from these two heat exchangers, may also be installed to preheat the cold tap water entering the water heater.

The air conditioning system employs an absorption chiller. This unit will require a larger generator surface and cooling tower than steam, gas, or oil-fired units. Because this absorption chiller will have a relatively small temperature drop, it may be feasible to use the return water from the chiller to supply hot tap water.

For steam district heating and cooling systems, the user equipment is similar, also, to that shown in Figure 1. The exception is that no heat exchanger will be required for the heating system if the building is currently steam heated. Steam systems may also require only a single pipe entrance to the building if there is no condensate return to the heating plant.

TABLE 2

Major Building Changes Required for District Heating

Existing system	District heating system	
	Hot water	Steam
Forced air	Replace furnace with heat exchanger	Replace furnace with heat exchanger
Hot water	Replace boiler with heat exchanger	Replace boiler with heat exchanger
Steam	Replace boiler with heat exchanger Replace valves Replace radiators	Replace boiler with PT Station
Baseboard electric	Replace baseboard units with radiators Install heat exchangers Install piping to radiators	Replace baseboard units with radiators Install piping to radiators Install PT Station
Heat pump	Replace heat pump with heat exchangers, absorption chiller, and cooling tower	Replace heat pump with heat exchangers absorption chiller, and cooling tower
Absorption chiller	Modification of chiller	None
Compression-type chiller	Replace compression chiller with absorption chiller Increase cooling tower capacity	Replace compression chiller with absorption chiller Increase cooling tower capacity
Compression-type central forced air	Replace compression unit with absorption chiller and cooling tower	Replace compression unit with absorption chiller and cooling tower
Compression-type window units	Remove window units Install absorption chiller and cooling tower Install piping system	Remove window units Install absorption chiller and cooling tower Install piping system

Substantial modifications to the existing heating or cooling system may be required, in addition to the changes outlined above. Table 2 shows these changes for both hot water and steam district heating and cooling systems. In addition to these changes, piping from the system's main user equipment is required.

Although it is technically feasible to retrofit most buildings with district heating, it is difficult to convince a building owner to tie into the district heating system and to replace an existing system. Even though the district heating system may be economic in the long run, based on fuel and operating costs, the cost savings are generally not sufficient to justify the capital cost of such a change. The only exception to the retrofit problem is a building which has a heating system due to be replaced. Since this problem is nearly universal and presents a serious threat to the implementation of district heating, a solution or number of solutions must be found if district heating is to grow in this country.

An application of techniques used in Europe, with some modifications, may be the best first step in solving the problem. The key to their success involves a three-part approach.

First, there is a near mandatory connection to the district heating system by government regulation when district heating is extended into the area. This approach has faced opposition in this country from the traditional social values of individual independence and reliance on the free enterprise system.

The second step is purchase of the existing heating system by the government through outright purchase or special tax treatments.

The third step is long term (30 years) government loans or government guaranteed loans help to reduce the need for, or opposition to, mandatory connection.

These last two steps amount to a government subsidy of district heating. It may be possible to phase these subsidies out as the district heating system grows, as the incremental cost of adding customers drops, and as the cost of fossil fuels rise. This last point is important, because as fuel and capital costs rise, the incremental increase in the cost of district heating should be less than the increase experienced by single-purpose individual systems. This is due to the greater fuel efficiency and economy of scale of the district heating system. Also, the district heating system would be based on coal-fired units, while the individual building systems are based on oil and natural gas. Since it is anticipated that the price of oil and natural gas will rise faster than the price of coal, while their availability will decline more rapidly than that of coal, the advantage of district heating over in-building systems is compounded.

Conversion of existing electric power plants to cogeneration plants serving a district heating system presents a number of technical problems, which in some cases may be impossible to overcome. However, the possibility of conversion must be investigated because of the capital cost advantages of conversion over new construction.

For a power plant to be considered for conversion, it must first be near the area to be served by the district heating system. This point is very critical for steam systems, since the maximum economic transport distance is approximately 3 miles. With a hot water system, the transport distance increases to as much as 70 miles, but the capital cost of installing such long piping systems cannot be justified unless the district heating system is large and well developed. Thus, the first plants which should be considered for conversion are the coal-fired urban power plants. These plants are generally smaller, older, and less efficient than plants which have been built far from the urban areas. Fortunately, these factors work in favor of their conversion to cogeneration. Since small, multiunit plants lend themselves to district heating applications from a system reliability standpoint better than large plants, and since their size more nearly matches the requirements of the district heating system, these plants are well suited for integration into the district heating system. As a cogeneration plant, they have a higher second law efficiency which makes them more economic to operate and, thus, a more valuable part of the electric generation system.

The second important consideration is the ability to obtain steam. If the turbine is a back-pressure design, then this poses no problem. However, most power plants use a condensing-type turbine from which steam must be extracted. This can be accomplished either by utilizing an existing extraction point, or by installing a new extraction line. Extraction steam should be drawn from the low pressure section of the turbine, where it is at a temperature and pressure closest to that used in a district heating system. This results in the greatest amount of work being done by the steam prior to extraction. A major problem which arises in retrofitting existing plants to cogeneration is maintaining proper steam conditions in the turbine. Excessive or improper extraction can lead to condensation in the low pressure section and turbine failure. A careful heat balance must be performed to assure proper turbine operation. Because of the limitations on the amount of extraction steam which can be taken, these plants cannot contribute all of their reject heat to the district heating system, as can a plant originally designed for cogeneration. This means a lower overall second law efficiency for the retrofitted plant when compared with the plant designed for cogeneration. However, there is a higher efficiency than there is without retrofitting. Because of the variations in plant and turbine designs, each plant must be evaluated individually. Can retrofitting for cogeneration contribute sufficient energy to the district heating system to justify the capital cost of the project?

An additional piece of equipment required in retrofitting is a pressure-temperature (P-T) reducing station for a steam district heating system which will bring the extraction steam conditions down to those of the district heating system. Depending on the extraction conditions, it may also be advantageous to use a P-T station on a hot water district heating system prior to the steam to water heat exchanger. Since the P-T station does reduce the useful energy available in the steam, it is most advantageous to match the extraction steam conditions as closely as possible to the district heating system conditions. This eliminates the need for pressure and temperature reductions. This design, however, must be within the base-load constraints of the plant if steam dispatch is based on the electrical system load.

VI. INSTITUTIONAL CONSTRAINTS

The wide-scale use of district heating in the U.S. has been hampered more by institutional constraints than by any technical considerations. One of the primary constraints on district heating systems has been the low cost of natural gas to the individual user. Because gas is a clean, convenient source of energy applicable to both central and domestic heating, the thermal efficiency advantages of central heating are greatly reduced. Continued government regulation of the price of natural gas is perpetuating this disadvantage.

The International District Heating Association has for many years promoted district heating in this country and aided in the exchange of information between district heating companies. However, they have not had a significant influence on governmental policies and the general public. There is presently no organization working effectively to promote district heating through legislation and public support.

There are no existing governmental measures which will insure full connection to a district heating system in an area in which it has been shown to be economically viable. Direct cooperation between the local government and the utility is necessary for district heating to be a viable alternative.

At the state level, the public service or public utility commission would need to be cognizant of the advantages of district heating, along with the costs of such a system and the requirements for right of way. In such circumstances, the utility has greater assurance of obtaining a fair rate of return and timely implementation of the system.

The existing air quality requirements imposed by the national, state, and local governments do not offset the potential improvements in overall air quality which can be achieved by implementing a central heating plant with flue-gas cleaning as compared to the existing emissions from multiple low-stack installations without flue-gas cleaning. District heating is most applicable in the densely populated urban areas where the air quality problems are the most severe, and the regulations are the strictest. These effects are most important in areas where coal and oil are the primary fuels for individual heating systems.

The electric utilities in this country are, for the most part, reluctant to venture into district heating or expand their existing systems. Many utilities have at one time had district heating systems, but abandoned them as costs of service rose and cheap natural gas become plentiful. Other utilities which still have systems have seen the number of customers and revenues dwindling. There is also a reluctance by utilities to build the small plants most applicable to district heating. These considerations have kept many utilities from reevaluating district heating systems which employ modern technology as a possible economically attractive alternative to other forms of space heating in their areas.

It appears that at least four social groups (the utility, the building owners, local and state government, and the environmentalists) must actively support district heating before it will achieve acceptance. While in some locations one or two of these groups are active, there are few, if any, locations where the interest has been generated within all four.

VII. ENVIRONMENTAL CONSIDERATIONS

Air and water quality requirements are of critical importance to the success of district heating systems. This is because most existing, and probably future, systems are, or will be, centered in urban areas where the control of pollutants is most critical.

Nearly all current in-building systems exhaust waste heat directly to the atmosphere. This produces widely dispersed, low-level temperature effects. The use of central cogeneration plants concentrates the discharge heat at a river, lake, or ocean in an open-cycle system. Although the amount of heat rejected to the environment is considerably less for the central plant of the same Btu input, the discharge is still a large point source which must be carefully controlled and monitored. About 35 to 45% of the energy normally lost from the in-building systems cannot be accurately measured or accounted for in the analysis of the district heating system. No credit can be taken for this saving, since the only measurable change may be a slight variation in local temperatures. From an air quality point of view, the measurable effects are quite different. In northern Europe, where fuel oil was the primary source of fuel for in-building systems where district heating was not available, the introduction of district heating has resulted in substantial reductions in SO_2 levels. Although the primary fuel for in-building systems in most parts of the U.S is natural gas, improvements in air quality are obtainable. In addition to reduction in SO_2 levels, improvements can be made in the concentration of CO, CO_2, and NO_x by using district heating. The use of a large central plant allows the installation of air-quality-control equipment which would not be economical, operable, maintainable, or practical on small, in-building systems.

District heating via cogeneration offers substantial reductions in reject heat to the atmosphere and improvements in air quality in the urban areas which suffer most from these problems.

VIII. FINANCING AND COST OF SERVICE

The success of district heating requires the ability to raise large amounts of capital and the acceptance of relatively low rates of return. Both of these requirements are found in the electric utility industry. To maintain the cost of service as low as possible, essentially all buildings in the service area must be connected to the district heating system. This implies a regulated monopoly, another condition met by the electric utility.

A large part of the financing of the system must come from sources providing long-term loans. However, the cost of these loans could be reduced with government guarantees. The rationale for the guarantees is that it is in the best national interest to reduce our dependence on natural gas and oil. Since the district heating system is an integral part of the community it serves, local bonding is another source of funds.

In Europe, subscriber loans account for a large fraction of the financial support for the district heating system. The subscriber loans the district heating company an amount equivalent to his alternative cost of an in-building system, but prorated on the age of the system. For residential customers, government loans are available. Electric utilities with existing electric plants which can be retrofitted for district heating can apply all external financing to the installation of the transmission and distribution network because the cost of plant retrofit is a small fraction of the cost of the system and can be paid for from internally generated funds. This also assures the customers and financiers that their loans are totally attributable to the district heating system.

TABLE 3

**Transmission and Distribution
System Cost Components**

Component	Cost (%)
Civil work	30—50
Changes in services	5—10
Piping	15—30
Pipe fittings	12—15
Insulation	1—4
Testing	5—6

The cost of service must be initially attractive, but also must remain attractive to assure that the customer remains on the system, and that the system will grow. In Europe, this problem does not arise since the State mandates, for competitive economics, that all buildings must be connected to the district heating system when it is introduced into the area. The fee can be structured such that there is a fixed fee based on the maximum annual contracted flow rate and a variable fee which is based on consumption. The fixed fee covers the capital-cost portion of the district heating system, while the variable fee is related to operating expenses.

IX. ECONOMIC ANALYSIS

There are four basic components which make up the capital cost of a district heating and cooling system: cogenerating plant, transmission system, distribution system, and in-building systems. The cogeneration plant will not be covered in the economic analysis because of the wide variety of possibilities for new and retrofitted plants, and because local regulations or corporate policy will determine the distribution of cost between the district heating and electric portions of the plant.

The transmission system is made up of large-diameter piping to carry the steam or hot water from the generating plant to the service area. As the transmission system passes into the service area, the smaller piping of the distribution system branches off of the transmission system to serve the customers. In general, the transmission system is made of long runs of piping with very few connections, none of which are connected to the customers. The exceptions to this are very large industrial customers who may be connected directly to the transmission system and provide their own in-plant distribution system.

The transmission system costs are made up of six major components as shown in Table 3. As can be seen by the percentages of costs, the civil work is the largest single component and the least certain. This uncertainty is the result of the dependence of the cost on size of pipe, soil and rock conditions, and stability and type of structure used (tunnel or concrete culvert). Both steam and hot water systems fall within the same cost range. Since there is no condensate return for the steam system, the cost of the higher-quality piping required for the steam system is offset by the use of only one line and by the smaller structure.

The distribution system costs are similar in makeup to those of the transmission system and suffer from the same uncertainties in the cost of each component. Smaller pipe diameters for the distribution system yield lower costs per foot of pipe laid than for the transmission system and allow the direct burying of insulated steel or plastic pipe, thus reducing the percentage of civil work in the total cost. The use of plastic pipe is new and should be considered experimental at this time. If its use proves prac-

TABLE 4

Cost Components for District Heating
and Cooling[2,3]

$/ft

Transmission	
Urban	750 ± 250
Suburban	500 ± 200
Rural	500 ± 200
Distribution	
Urban	300 ± 200
Near Urban to Suburban	200 ± 500

Commercial Buildings

Conversion from	Steam ($/kWh)	Hot water ($/kWh)
Steam boilers	2 ± 2	25 ± 5
Hot water boilers	25 ± 5	25 ± 5
Roof-mounted forced air	75 ± 5	75 ± 5
Absorption chiller	20 ± 10	25 ± 10
Central compression cooler	75 ± 15	75 ± 15
Distributed compression cooler	>300	>300

tical, its lower cost may make it economically attractive for use in single-family residential areas.

Two primary parameters are employed in determining the economics of district heating systems. A cost per foot ($/ft) calculation is useful in evaluating the alternative piping schemes, types, and installation. A cost per unit of heating capacity ($/kWth) calculation reflects the capital cost which must be recovered from the user, and thus, will determine the most economically attractive areas for introduction of district heating and cooling.

The conversion of in-building heating and cooling systems is usually expressed in $/kWth as shown in Table 4. In general, the costs for commercial buildings are inversely proportional to the size of the system. Industrial plant conversion costs are not easily estimated because of the difficulty in defining general costs for such a diverse category. However, nearly all industrial applications have existing in-plant steam systems for process heat (not hot water), and conversion will mainly entail matching the pressure level of the internal system with the pressure in the district heating system.

It is extremely difficult to estimate the cost of conversion of existing buildings to district heating and cooling without knowing the design of the building and the existing HVAC system. The trend in system designs in recent years for commercial buildings in the U.S. has been to either hot water boilers and absorption chillers supplying floor or zone forced-air heat exchangers, and to roof-mounted forced-air heaters with compressor-type air conditioning. These systems apply to high-rise and low-rise buildings, respectively. European experience is of only limited help in developing conversion costs because most predistrict heating systems were oil- or coal-fired hot water boilers, or electric heating. The insulation and window glazing requirements have been higher than in the U.S. In addition, there is little or no air conditioning load. The variety of heating and cooling systems which may be encountered in converting existing buildings to district heating and cooling were discussed previously (Table 1). As might be expected, the more difficult and involved the conversion, the greater the cost to the building owner.

Transmission system costs are usually expressed in terms of $/ft of length. This provides a cost of installation based on the difficulty of installation and the pipe size, as shown in Table 4.

Distribution system costs, on the other hand, are expressed in terms of both $/ft and $/kWth, as shown in Table 4. The first method yields the same information as in the transmission system calculation. Unfortunately, this does not adequately reflect the effect of the load density in the service areas. Using the $/ft calculations and applying the load density (kWth/mi²) yields the $/kWth figure, which is a good reflection of the effect load density has on system costs. It also reflects the distribution of capital cost to the customer.

REFERENCES

1. **Larsson, Kjell,** District heating: Swedish experience of an energy efficient concept, unpublished data, 1978.
2. **Oliker, I. and Philipp, J.,** Technical and economic aspects of district heating systems supplied from cogeneration power plants, paper presented at the Am. Power Conf., Chicago, Illinois, April 24 to 26, 1978.
3. **Karkheck, J., Powell, J., and Beardsworth, E.,** Prospects for district heating in the United States, *Science,* 195, 948, 1977.

Chapter 11

TECHNICAL AND ECONOMIC FACTORS FOR INDIVIDUAL INDUSTRIAL PLANT SYSTEMS

William S. Butler and Paul Hodiak

TABLE OF CONTENTS

I. INTRODUCTION

In this chapter, cogeneration in large industrial plants is treated separately from cogeneration in small industrial (or military) installations. In general, large plants will find it both technically and economically more feasible to install steam turbine systems, to use coal-fired boilers, and if considered beneficial, to operate fully independent from the local utility. Although large industrial plants could install diesel engine or combustion turbine systems, they would then for the most part be faced with a need to sell large amounts of surplus electricity. Conversely, as is shown in the second part of this chapter, gas turbine or diesel cogeneration may be the preferred technology for the smaller plant.

II. TECHNICAL AND ECONOMIC CONSIDERATIONS FOR IMPLEMENTING COGENERATION IN LARGE INDUSTRIAL PLANTS*

A. Isolated Electrical Operation

A preliminary study will indicate if the quantity of electric power which can be economically generated on the steam base will meet the needs of the plant (see Chapter 1, Figures 11 and 13). If there will always be a slight surplus, consideration can be given to cogenerating the entire plant electric load and operating with no interconnection to the utility (isolated operation). Two factors must be considered. First, *how critical is continuous service* to the electric power load? Second, *how reliable is the generating plant?* If any considerable part (say, 25%) or more) of the electric load can be considered as "interruptible", and if the generating capacity is in several separate units, this system may be the most attractive. A very high degree of reliability (typically 98%) is needed in the power plant. It must be designed to be self starting, to be capable of subdividing units for operation, and it should be conservatively rated. The electrical load should be distributed in several feeders arranged by priority of load and equipped with an automatic load-shedding capability. If the quantity of power generated cannot meet the entire demand, some of the load may be segregated and be permanently connected to the utility. This would usually be the highest priority load to avoid the high utility stand-by charges on such service. The remainder of the electric load is carried by the cogenerating power plant in isolated operation. The voltage and electrical frequency on this portion of the load will be controlled by the cogenerator. Any critical part of the electric load can be equipped with a double throw switch or with mechanically interlocked circuit breakers so that it can be served by either the utility or the cogenerator. The standby charges which the utility must make for this type of service are usually very high, since the potential demand must be covered by "rolling reserve" capacity maintained continuously in service by the utility. This type of cogeneration is not particularly beneficial to the utilities. However, it may result in substantial savings to the cogenerator if the plant heat-and-power energy match is favorable, and the loads are relatively stable.

B. Electrical Interconnections with the Utility System

In many cases, large industrial plants will have waste heat, waste fuels, or large heat uses that permit surplus electric power to be generated. In these cases, it may be advantageous to arrange a synchronous interconnection with the electric utility and sell the surplus power. Since it will usually require only 4400 to 5000 Btu/kWh for such

---

* The following material is largely drawn from the working experience of the writer and is not elsewhere documented for reference.

power to be generated, it most likely can be produced at a lower cost than the utility would incur.

Other factors favoring a synchronous interconnection are high priority electric loads or relatively few individual power units in the cogenerators plant. In the following discussion, several necessary attributes of the interconnected cogeneration system are reviewed.

C. Reliability Considerations

Assuming that the preliminary study has indicated the economic viability of surplus electric power generation, reliability becomes the prime consideration. Many otherwise viable cogeneration projects have failed where their performance was not sufficiently reliable. *No power generating equipment is capable of 100% availability).* If spare equipment is too expensive, the alternative is to provide a synchronous interconnection to the utility. However, if this interconnection does not offer advantages to the utility to compensate for the additional liabilities incurred, the utility cannot favor it. Through interconnections, there can be mutual advantages to the industrial plant and the utility. In order to obtain them, the industrial cogenerator must adopt the basic philosophy which motivates *successful* utility operations; *uninterrupted service.* Consider that the utility is charged with a demanding responsibility for the safety of its line employees and the general public in connection with its transmission and distribution facilities. Every generating plant on the system, however small, introduces another hazard which must be guarded against when undertaking repair work. It is imperative that the utility have total control of the interconnection. They will need control of the power transferred, the direction of flow, the voltage, and the frequency. Relay protection of the interconnection must be under control of the utility. The switching and transmission line rules must be those of the utility, as must be the isolation and safety system.

For the cogenerator, there are advantages to these requirements. If the standards of operation and maintenance of the utility are observed, the standard of reliability for the cogenerator and the nearby customers of the utility will usually be much better than either could achieve separately. An example of such an interconnection occured when the Ludington Power Plant of the Dow Chemical Co., which had previously operated isolated, interconnected with Consumers Power Co. through the Michigan Public Service Co. Based upon the number of interruptions to electric service per year, the City of Ludington dropped to 20% of their previous frequency, while the Dow plant dropped to one third of the interruptions experienced with isolated operation. The utility purchased surplus power on a peak-load schedule at a favorable rate. The capital for a new transmission line into the Ludington area was delayed for many years. All operations were controlled by the utility. During the few occasions when stand-by power was required, the watt/hour meters were allowed to run backwards, but always on a *pre-arranged basis.* On such occasions, generators were continued in service for voltage control even when there was insufficient steam to generate significant power. Both parties to this agreement benefitted to the extent that it continued for 18 years without material change. Although the contract was for "firm power", which required Dow to shed internal load in order to save the utility service, this was rarely necessary. Far more frequently, coal-feeding difficulties in the cogeneration plant forced brief reductions of power to the utility. It was only necessary to report these as soon as possible to the utility power-control center.

The typical cogenerator is a relatively short tail, and the tail cannot expect to wag the dog. With this philosophy in mind, we can identify other characteristics of a cogenerating plant which should facilitate interconnection with the utility system.

D. Voltage Control

The standard for industrial size, air-cooled electrical generators is for 80% power factor. This level requires no additional capital. However, for interconnection, the cogenerator's electrical system should be designed for at least a 90% power factor by the use of capacitors or a few synchronous motors. This practice is economical. It will cost the industrial plant very little to carry this additional "wattless' or reactive current, but it may materially assist the utility in maintaining proper voltage during critical periods. If this capability is predictably available to the utility, it can be particularly advantageous. It is often possible to supply standby power if the cogenerator is willing to load his generators with "wattless" current during boiler outages. This practice requires scarcely any extra fuel.

E. Surplus Capacity

If there is inventory or surge capacity in any industrial process so that high process-heat rates can be absorbed for a portion of time, the cogenerator's turbines may operate at higher rates during peak-load hours. In this case the power is far more valuable to the utility than base-load power would be. In many cases, common prudence in design would dictate building in such surplus capacity. A willingness to use it to produce maximum peak-load power for the utility could be a valuable asset.

F. Firm Power

Here we must consider the characteristic denoted in utility circles by the word "firm" applied to electric power, because the degree of "firmness' determines the value. Firm power means dependable power. Thus, the power from a cogenerator's noncondensing turbine exhausting steam to a process could only be considered as firm, if power output would be maintained by exhausting steam to the atmosphere should the process suddenly shut down. Thus, if the turbine was normally operated under *automatic* back-pressure control, it would be difficult to sell the resulting power as firm. If, on the other hand, the turbine was operating on speed-governor control, a drop in frequency would cause it to pick up load even if the steam process went down simultaneously (by blowing steam through the atmospheric relief valves). This power might then be considered firm, provided atmospheric exhausting would be continued until permission from the utility to cut load was obtained.

G. Turbine Selection

A straight condensing turbine would be economical to a cogenerator only if waste heat or very cheap waste fuel was available. It might also be a valuable source of peak power if the waste fuel could be stored. A more attractive approach for industrial cogeneration is the automatic extraction condensing turbine which can produce low pressure process steam at up to three different pressures, with only a very small quantity entering the condenser. Such units can quickly pick up a full electrical load even if all exhaust-steam uses are briefly lost. Unused capacity on this system could be considered as "firm, rolling reserve" if subject to utility need. Peak-load-period operation of this extra condenser-turbine-boiler capacity could be valuable at times to the industrial plant.

H. Standby Power

The approach to obtaining the most economical standby power from the utility is to arrange scheduled plant shutdowns with the utility power-control department so as to take their power on a base-load, around-the-clock schedule. When the standby power is no longer needed, arrangements might be made to pay it back a little at a time during such peak-load hours as they may designate. In any case, this will usually provide the most reliable and inexpensive standby power.

I. Cranking Power

Start-up assistance is another service which a cogenerator can provide, and which, under certain circumstances, can be very valuable to the utility. Most utility power plants are incapable of unaided start-up. The great Northeast blackout of recent history hinged sharply on this problem. The cost of retrofitting utility plants for unaided start-up would be prohibitive. Many schemes for obtaining cranking power have been considered. It costs relatively little to design an industrial power plant to perform this function, since steam-turbine-driven auxiliaries are usually easily justified. The electric cranking power requirement becomes very small. All cogenerating plants should be so designed. A cogenerator with all electric auxiliaries cannot provide the magnetizing current to energize a transmission line without knocking out his own auxiliaries. If the auxiliaries are steam turbine driven, the generator capacity is the only limit. Since the surge is brief, much excess can be tolerated, provided that the voltage dip is not critical. The generator field voltage can even be applied gradually. Furthermore, steam-driven auxiliaries are relatively unsusceptible to lightning or other line disturbances. Note that the conventional alternating current rectifier exciter cannot be used unless it is supplied by a *separate* main-turbine-shaft-driven alternating current generator.

J. Feeder Design

The arrangement of the electric load in the cogenerator's power system may also enhance the value of the interconnection to the utility. If possible, the load should be distributed among several feeders, keeping the high-priority loads separate from the low, and establishing the order for dropping feeders when necessary so as to minimize the loss to the industrial plant. Assuming that the low-priority load can be occasionally dropped without significant harm, the value to the utility is obvious. Although such occasions will be rare, the cogenerator should be prepared to drop such load to protect the voltage and frequency of the interconnection. This makes for increased security of the high priority loads at all times. An automatic load-shedding device can readily be applied for further protection of high-priority circuits.

A cogenerating plant should be equipped to control line frequency and maintain accurate time during periods of separation from the utility system. It should also have synchronizing equipment so that it can return to the system without interruption to service or causing surges.

K. Steam Boiler Considerations

1. Package Boilers

For a number of years, the majority of new industrial boilers have been shop-assembled packaged units. They are light, compact, cheap, and convenient. They are usually designed to burn only natural gas or light oils. They will not tolerate low quality feed water since they normally operate at very high ratings and have relatively small, thin tubes. A major disadvantage of these boilers is that they cannot be converted successfully to solid fuels. It is also difficult to make an effective conversion to heavy fuel oils. Such boilers are generally unsuited to cogeneration projects. However, in some cases they could continue to serve as a standby steam source. There is also a possibility of their use (with a mixed-pressure turbine) to keep some generation in service during coal-fired boiler outages, for peak-load shaving, and for start-up or voltage boosting. In any case, where cogeneration equipment is replacing package boilers, these units should be retained until their continuing value can be assessed.

2. Types of Boilers

If the potential for industrial cogeneration is to be fully exploited, a large percentage of existing industrial boilers will require replacement. From fuel availability and economic considerations, it is expected that the majority of these future boilers will be both high pressure and coal fired. The capacity of boiler manufacturers to meet this requirement may be strained. Currently unproven newer products will not supply the need for many years. A substantial return to past practice may smooth the transition. Some older boiler designs were quite successful and, perhaps, should be resurrected. A few general remarks might aid in passing through this cycle again without repeating past errors:

1. To be efficient, cogeneration requires high steam pressures, 600 psig or over. Only water tube boilers can be considered. Fire tube boilers are usually limited to sizes under 20,000 lb/hr and to steam pressures of 200 psig and are, therefore, not appropriate.*

2. Industrial water tube boilers can readily meet the size and pressure requirements. The bent tube designs are generally superior to straight tube designs for ease of tube replacement, better circulation, and flexible configurations to fit fuel requirements. Bent tube designs employ far fewer stays and hand-hole plates. Some straight tube designs have as many as two hand-hole plates for each tube. High pressure boilers should avoid stay bolts which were once quite common in straight tube boilers. Repair of broken or cracked stay bolts have been a common cause of much extra expense and downtime. Both hand-hole plates and stay bolts will often develop leaks following chemical cleaning operations.

3. Boiler designs employing extensive refractory material in the furnace are not recommended. Air-cooled refractory walls are preferred to solid. Water-cooled furnace walls, burners, and even water-cooled stokers such as the "vibragrate" tend to reduce maintenance costs. However, avoiding *all* refractory, particularly around pulverized-coal burners, may lead to serious flame-instability problems and possible furnace explosion. One can, though, enjoy some of the advantages of both worlds by covering the water-cooled burner tubes with spot-welded studs to anchor the layer of refractory necessary to promote good flame propagation.

4. Adequate furnace volume is always a prime consideration. Older designs were often deficient and, if used today, would be confined to prime coals. An undersize furnace will prevent the use of western coals, will promote excessive slagging problems, and will likely be unacceptable from environmental considerations. Ozides of nitrogen (NO_x) readily form at the excessive furnace temperatures therein.

5. Certain types of stokers exhibit a tendency to smoke. Of these, the most troublesome is the spreader stoker firing upon a stationary dumping, pinhole grate. The larger sizes employing a front-moving traveling grate do not require the grate dumping and rekindling operations which tend to produce smoke. The traveling-grate design continuously cleans the fire by dumping the ash over the front of the grate into the ash hopper. This action can be fine tuned to be nearly free of smoke and cinder carry over. All spreader stokers require effective cinder catchers together with a system of refiring the cinders into the furnace, since they contain large quantities of unburned carbon.

---

* When feed water is of poor quality, fine tube boilers can be built up to 70,000 pounds steam per hour and 200 pounds W.S.P.

L. Operational Considerations

Stokers on coal-fired boilers can waste large quantities of fuel if carelessly operated. A rotograte spreader stoker operating with a deficiency of air can waste up to 20% of the fuel by incomplete combustion of furnace gases and can also simultaneously lose another 20% by dumping unburned coke into the ash pit, meanwhile making dense smoke and causing damage.

Pulverized-coal-fired furnaces behave differently. Either insufficiency or a serious excess of air which could cause excessive losses will usually impair the flame propagation and cause the fires to go out. This in turn will cause a master fuel "trip" and a boiler shut down. Either system can damage the boiler through secondary combustion. Pulverized coal is also subject to the hazards of furnace and pulverizer explosions. If older pulverized-coal systems are to be resurrected, prudence demands that these be fitted with modern electronic furnace safeguard systems.

1. Economic Superiority of Pulverized Coal

Since stokers are limited to 300°F preheated air temperatures, this requires the use of a small air preheater and a very large economizer in order to secure high boiler efficiency. The large economizer must be supplied with relatively cool feed water (not over 250°F) if steaming in the economizer is to be avoided. Steaming economizer designs are generally expensive and difficult to justify. This precludes the use of a regenerative-cycle stage heater using extracted steam of over 25 psi. A pulverized-coal-fired installation can use hot air of any temperature and will often require temperatures of about 600°F in order to dry wet coal. Thus, the air-preheater surface (which is relatively inexpensive) can be very large, and the more expensive economizer surface can be quite small. It is then feasible to install high pressure regenerative-cycle feed-water heaters for water temperatures up to 450°F. With such an arrangement, and even allowing for higher auxiliary power requirements, the thermal efficiency using pulverized-coal firing can be several percentage points better than stoker firing.

M. Future Considerations

The future for industrial coal-fired systems is almost completely dependent upon regulations promulgated by the Environmental Protection Agency. If these are not too stringent, fluidized-bed coal combustion appears to have great promise. Both intermediate and small boiler units, operating at atmospheric pressure, have been successfully demonstrated, and more than one vendor is offering to build and install such units in industrial plants.

III. TECHNICAL AND ECONOMIC CONSIDERATIONS FOR IMPLEMENTING COGENERATION IN SMALL INDUSTRIAL OR COMMERCIAL INSTALLATIONS

This part of Chapter 11 will cover the implementation details of four small industrial/military cogeneration operations that have been designed and installed by the writer and his associates at Applied Energy, Inc.

A. General Characteristics

Each steam plant covered in this discussion has its individual design. No two plants, at this time, have the same design. The cogeneration facilities are all gas-combustion turbines driving electric generators. The electricity produced is placed on the utility grid. The turbine is equipped with a heat-recovery boiler. Each boiler is sized to the individual turbine supplying the exhaust gases. The turbine gases enter at peak temperatures of 970 to 850°F. The electric outputs for the four plants (A, B, C, and D) are 14.7 MW, 27 MW, 22 MW, and 0.8 MW. The heat recovery units produce 119 Mlb, 115 Mlb, 127 Mlb, and 7 Mlb of steam per hour, respectively. The feed water to each boiler passes through sodium zeolyte softeners. In one of the four, a split-stream de-alkylizer unit employs sodium zeolyte and hydrogen zeolyte softeners and neutralizing amines. A deaerator and blowdown heat exchanger are also employed at each installation.

B. Operating Performance

The general plan for the turbine is to maintain a dual-fuel capability. Oil can be easily burned, but gas feed requires high pressure gas. The unavailability of high pressure gas in some areas imposes a restriction on the use of this fuel. In this region (California) all applications employing turbines (or even a package boiler) must meet, and continue to meet once built, stringent air quality standards.

The single most difficult problem that has to be dealt with in attempting to build a new AEI-designed cogeneration plant is the local air quality standards. Emission trade-offs are expected to be required regardless of increased, decreased, or unchanged steam loads or fuel requirements. Similarly, the single most demanding operating problem is maintaining the turbine within the air quality standards once it has been built.

The stringency of emissions control will generally be inversely related to the turbine heat rate, which translates directly into operating economics. If the reader is in an area of the country that does not have stringent air quality enforcement, there may be a feeling of over emphasis in this category. However, fuel costs represent 70% of operating cost, and project viability necessitates control of this cost exposure. Emission rules also apply to the standby package boilers (see Chapter 3) and can limit their maximum output, thus creating operating and maintenance restrictions which were unthought of during the formulating, contracting, and building of these plants.

At plants B and C with the larger turbine facilities, the estimate of increased heat rate for NO_x suppression is 500 to 750 Btu/kWh. The cost impact, per turbine, will run between $250,000 to $350,000/year.

Plant D, which is designed to be an unmanned facility, is the smallest size cogeneration facility still having attractive economics, and the largest size acceptable for quick approval by the local air quality authority. From an air quality perspective in California, a maximum electrical capacity of 1 MW currently appears to be the limiting restriction for placing a new cogeneration plant into service in place of existing generating equipment. EPA New Source Emission Standards apply above this limit so that retrofit of existing steam operations becomes a much more sizeable capital investment.

There is a secondary problem related to the emission control of base-load combustion turbines. Some turbines do not have a sufficient history of operating in this mode to give the facility designer all the information needed for the original design. Specifically, for some equipment it is not possible to determine what downtime for maintenance will be incurred under NO_x control and base load. The data are inadequate for alternate system comparisons. Facilities in plants A, B, and C have run in the range of 8100 hr/year down to 5200 hr/year. Comparable time on comparable machines with comparable operations do not presently exist.

Plant performance can be measured in many ways. From an operator viewpoint, it is important to make sure there is enough instrumentation to allow a complete verification of the plant operating efficiency. Personal experience with large and small industrial and commercial customers has shown that the initial capital expense for the performance-monitoring instrumentation is often eliminated as a "noneconomic" use of capital. However, cogeneration plants are almost useless without this monitoring capability. It is vital to know where the plant stands relative to energy use and cost factors at all times. In either large or small plants, cost effective operation requires this over view. The smaller the cogeneration facility, the more critical this instrumentation becomes, and yet, the greater the pressure is to eliminate it as a capital expense and an on-going maintenance expense. It is expected that as fuel prices climb, the cost impact of loss of thermal performance will become increasingly significant and, thereby, more readily justify the required instrumentation.

The heat-recovery boilers in these plants nominally recoup 30 to 35% of the turbine thermal daily input during the heating season. High performance days of 40% recovery have been recorded, but those are infrequent with a diversified, nonprocess load. The 30% value has been achieved on a consistent basis regardless of winter load swings. The turbines are mainly operated in a base-load electric generation mode with a stack damper spilling excess thermal energy. The alternate electrical (MW) load-following/ steam demand feature (see Chapter 4) is installed on plants C and D, but has not been used sufficiently as yet to report its capabilities in following the daily load swings.

C. Operating Procedures and Instructions

Each cogeneration plant is operated by personnel from the local utility. A series of procedures have been developed and put in writing to cover both normal and abnormal operations of the plant.

Each manned facility is covered 7 days/week, 3 shifts/day, one man per shift, with a five-man shift crew of qualified plant operators and two relief operators. Manning requirements over the years have shown that an average 1.1 men are required on each shift to handle the operations. New facilities will average 1.2 men per shift during the first year because of the various testing and maintenance needs. The primary responsibility of the plant operator is to maintain the contracted continuous steam supply in a safe manner. The conditions of the equipment are continuously monitored in the central steam plant's operation control room. As indicated by the fractional additional man per shift requirement, there are times when a manual operation requires more than one person. During the greater part of the time, maintenance people are in the area of the plant to meet this need. An average of three maintenance people per day work on or visit a plant. A preplanned work schedule dictates when these people are at a given plant. There are two separate maintenance crews for the cogeneration plants. One is electrical, the other is steam.

Plant D is unmanned, but it is expected that operating and maintenance crews will be on site about 30% of the time. Personnel from the industrial user's operation will also assist in the inspections. Operating procedures are prepared and listed at the site.

The one primary concern in the operation of each plant is *continuous steam production*, which is the contractual obligation. Because of reliability concerns, an industrial plant will not usually contract for their steam supply outside the company. Thus, with AEI as an outside contractor supplying their steam, written procedures covering all contingencies are as great a help to the steam customer as to the operators. It is the one method by which the customer can feel assured of reliable steam plant operation. Additionally, the procedures for equipment failure, loss of pressure and supply, and for major-minor disturbances have to be approved by and coordinated with the customer. It is his property, distribution lines, and business that is impacted. Even if an outside contractor, such as AEI, were not used, it is still prudent operation to have the procedures and contingencies well documented.

AEI uses turbine exhaust-heat recovery as the primary source of steam. Each system occasionally requires supplemental steam at high steam demands. It is estimated that only 90 to 95% of the present hourly steam loads are within the maximum capacity of the heat-recovery boiler. To meet peak demands, the standby package boilers are brought on-line. Conversely, at low steam demands, the package boiler is used because the heat-recovery boiler is not economical. This is a consequence of the reduced efficiency of the turbine at low electrical output or at high rates of spilling of the turbine exhaust heat. A regular updating of the operating procedure is provided to the operators to inform them of the current heat-recovery-boiler lower economic limit for steam production. As later explained, the fuel cost of steam production is quite variable, and various components of steam cost will change at least monthly. This necessitates recalculating the lower economic operating point at least once per month. A different economic operating point applies to each cogenerating plant, and a separate procedure is written for each.

D. Fuel Costs for Steam Production

Due to the unique way thse projects are structured and operated (see Chapter 3), turbine fuel is not purchased by AEI, but rather, waste heat in the exhaust gases is. The pricing mechanism is an equitable way for anyone to view the cost of generating steam and electricity.

The cost of heat for producing the steam in the heat-recovery steam system is the difference between the cost of fuel the turbine consumes and the value of the electricity the turbine-generator set produces. The cost of fuel consumed in the turbine will be the combined cost of the natural gas used and/or the cost of the fuel oil burned. The value of the electricity may be more difficult to develop. For AEI, it is the average *fuel* cost of producing electricity by the local utility. For an industrial customer, the value of cogenerated electricity could be the cost of purchasing the equivalent amount of electricity from the local utility. All stand-by charges, facility charges, time-of-use charges or adjustments, and other possible costs should be reflected in this cost. There are many ways of accounting for the allocation of costs between electricity and steam production. The way AEI performs this allocation is of interest merely as an example of one specific application.

Figures 1 and 2 are fuel-cost graphs for customer sites A and B, which show the trend for lower fuel cost as net steam production increases. The straight lines show the trend, but the actual data points have a wide variability, due primarily to differences in daily load profiles. The examples represent actual daily results and show how they can vary for the same net steam production. Significant cost impact arises from the amount of package-boiler steam required. The more the heat-recovery boiler is used to meet the load demand, the less expensive becomes the steam. It is possible with a relatively high and smooth steam-load profile to generate steam at a fuel cost less than the turbine fuel-input cost.

Each graph charts the net steam-production fuel cost (turbine exhaust heat cost plus package-boiler fuel cost) against the net daily steam production. The scale on the ordinant is the ratio of cost per million Btu of steam to the delivered cost per million Btu of the turbine fuel. At the point 1.0 on the ordinate scale, steam is being produced at a cost equivalent to 100% fuel efficiency.

Figure 3 shows the turbine daily electrical (MW) output vs. the steam produced by the heat-recovery boiler. The variation of the data points from the trend line shows the impact of damper spill setting as well as other variables. The data points on the extreme upper right of the trend line are representative of the design maximum steam output of the heat-recovery boiler.

The graphs are based on observed daily operating data and portray actual costs. They are not engineering figures which subsequently may be corrected to reflect actual operating points or conditions.

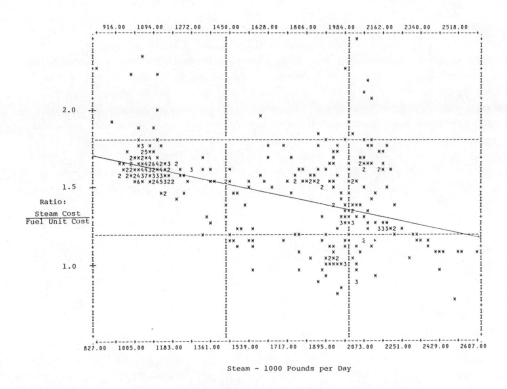

FIGURE 1.   Simulation of steam cost vs. steam production for Site A.

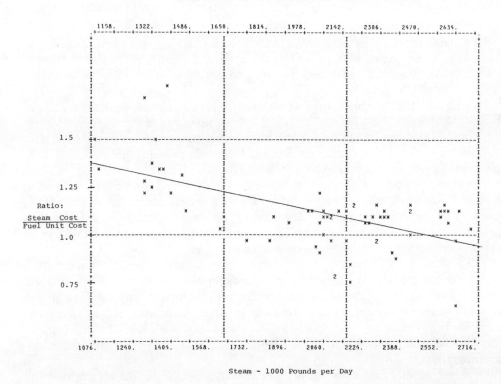

FIGURE 2.   Simulation of steam cost vs. steam production for Site B.

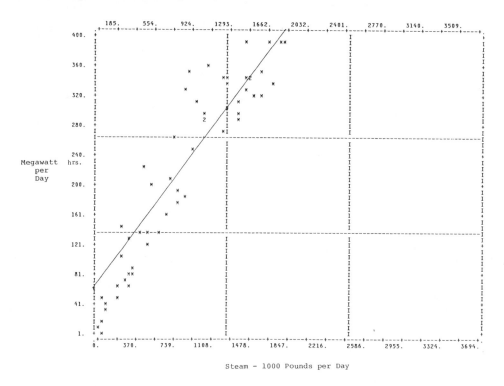

FIGURE 3.   Simulation of electrical power output vs. steam production.

E. Maintenance Costs

AEI maintenance policy changes frequently, as it tends to reflect management philosophy as well as economics. Reflecting the present preventive maintenance philosophy, expenses are uniform for each facility and unrelated to age of plant or annual usage.

Annual overhauls are performed on all steam as well as electric generation parts or systems. Equipment outages rarely force a system shutdown. Most equipment problems are handled on straight time, rarely necessitating overtime or immediate action. Redundant system design, including the dual fuel facilities and standby package boilers, provides this generally high system reliability.

The electric generation turbines are available a peak 97% during a high steam-load month. A low availability would be 70%. Throwing turbine blades has resulted in the loss of a turbine for a maximum 7 day period. The longest scheduled turbine maintenance period now performed under peak steam loads is 4 days (all weekdays). In all cases, operations strive for 100% steam availability.

The availability of redundant systems and prevailing maintenance philosophy will dictate the plant design. In plant D, little redundancy exists in the heat-recovery steam design. However, the user's standby boilers can easily take the steam load if the turbine or heat-recovery boiler requires maintenance. In this case, the plant reliability requirement has been reduced by the existing plant's readiness to maintain a backup capability. Even so, a relatively high reliability will be obtained through the scheduled maintenance program.

Qualified maintenance labor in this area is readily available. The high availability of the equipment is a direct measure of their effectiveness. The package boilers are available almost 100% of the time. Forced outages may amount to 1 to 2 days/year, but then only 1 year out of 2. Scheduled outages for maintenance last about 10 days maximum. The package boilers are taken out of service for this maintenance in the spring and fall while heat-recovery steam loads are well above the minimum economic, but well below expected peak demand. At these times, if the package boiler at any site is out of service, the heat-recovery boiler can serve as backup.

At present, the scheduled steam-system maintenance cost is equivalent to 2.5 persons per steam plant per year. In addition to the steam maintenance, an additional 3.5 persons are required to service the turbine. AEI is responsible for the nonfuel incremental operating and maintenance costs of the turbine. Experience has shown that the nonfuel O and M expenses on these turbines are comparable to those of fossil-fueled steam electric-generating plants. Such costs vary on a 3-year cycle, with 2 years of lower cost per megawatt hour ($/MWh). The 3rd year has higher cost. Turbine blades are replaced on a planned 3-year schedule. The nonfuel O and M cost is less than one half the cost ($/MWh) for the local utility peaking units regardless of maintenance activity.

Chapter 12

TECHNICAL AND ECONOMIC PARAMETERS FOR COMMUNITY SYSTEMS

W. R. Mixon

TABLE OF CONTENTS

I. INTRODUCTION

There are many feasible combinations of components and subsystems which could be used with the Total Energy (TE) concept to provide electric, space and water heating, and space-cooling services for buildings and communities. The number of options multiplies rapidly as the concept is expanded to include other utility services, as in the Modular Integrated Utility System (MIUS) concept, and other modes of operation that embody some form of connection with the local utility grid. Thus, community system design depends on many local, site-specific parameters. The manufacturers have, of course, developed standard product lines of major components with some attention to matching connecting hardware for TE applications, and there is a potential for increased modularization. However, community system design is largely custom tailored to each application.

Cogeneration conserves energy by increasing the utilization of input fuel energy. With gas-fueled internal combustion piston engines, about 32% of fuel energy is converted into shaft work when operating at rated load. The remaining fuel energy is rejected as heat: about 28% in exhaust gases, 23% to water-jacket coolant, 2% to the oil cooler, and the rest as radiation and convective losses.[1-3] Shaft horsepower may be used to generate electricity or drive other equipment, but improvement of overall system efficiency requires that part of the rejected heat be recovered and put to useful purposes. In typical community applications, essentially all of the water-jacket heat and about 60% of exhaust heat are economically recoverable. Further heat recovery, while possible, depends on site-specific requirements and economics. Thus, overall efficiency of the cogeneration system can reach 71% or more as compared to values of 30 to 35% for modern utility generating plants with transmission losses.[4]

It is this increased utilization of fuel energy and the corresponding decrease of energy purchases (as fuel or electricity) that enables cogeneration to economically compete with conventional utilities in certain applications. There are several prime factors that affect the economics of community systems and that illustrate the importance of community and system characteristics. A high fraction of recoverable thermal energy must be put to some practical use, i.e., to provide services that generate revenues for the plant and that usually would otherwise require consumption of utility electricity or fuel. Thus, recovered heat must be available in appropriate quantity at the times it is needed. The designer needs to consider ways to optimize the match of supply and demand for electric and thermal energy. Other important factors include (1) the plant load factor and (2) the relationship between cogeneration plant fuel cost and the cost of conventionally purchased energy. In practice, there will likely be a combination of favorable and unfavorable factors to be evaluated. Acceptable combinations have not been defined, but there are a number of basic requirements and system design practices that will be identified in this chapter.

II. COMMUNITY CHARACTERISTICS

Much of the information required to design and evaluate a community system stems from the location and physical characteristics of the community, such as:

1.   The variation with time of electric and thermal loads and other utility service demands on the plant
2.   The cost of distributing thermal energy
3.   The availability and cost of fuel
4.   The type of conventional end-use systems that would otherwise be used within buildings

5.   The availability and cost of each alternative conventional utility service

In some areas, new information or greater detail is required. In others, it is a matter of centralizing a design process that is conventionally distributed between developers, builders, subcontractors, municipalities, and utility companies. Community characteristics of interest are those that affect the design, energy conservation, or economic viability of community system projects.

A. Physical Characteristics

Included within a community system is the means to distribute thermal energy to the buildings to be served. When the community consists of a number of buildings, this subsystem may be considered as a type of district heating with the same dependence on the physical characteristics of the community.

An important aspect is whether the community is new, developing, or established. Installation costs for hot- and chilled-water piping systems will be the least for new communities in which system design and installation can be completed in phase with other land improvements. There are significant additional cost burdens and community disruptions where existing sidewalks or streets must be excavated and repaired and where there is an existing utility infrastructure, which probably is poorly documented, to contend with. These factors, together with the class of soil, whether common or separate trenching[5] is used, and local cost rates, affect earthwork costs for installing underground conduit. Installation costs vary widely between given sites, but for initial estimating purposes under average conditions, allowances of 50% of total material costs for installation in open country and 300% for installation in a city area have been suggested.[6] These factors are applicable to systems used for medium and high temperature water conveyance (up to $\sim 215°C$) in conduits that consist of a steel core pipe, insulation, and an outside casing. Different, larger cost factors would apply to less expensive conduit systems used with low temperature hot water and chilled water because earthwork costs are not sensitive to the conduit materials used.

It is generally recognized that the installation cost of underground conduit in urban areas where there is interference from existing utility lines can be 4 to 6 times the installation cost in open country. However, such generalized cost factors should not be used without some knowledge of site conditions. On a particular city street, the conduit may be installed under a sidewalk with a relatively small extra cost and costly interference conditions may only exist where the conduit crosses streets or intersections. Thus, high installation costs attributed to urban areas may actually apply to only a small fraction of the total length of conduit to be installed.

The status of the community also has an important effect on the cost of end-use HVAC systems within buildings. The planned use of hydronic systems within new buildings may offer a cost advantage, but the cost of retrofitting some existing buildings may be prohibitive. Both the type and age of existing equipment should be considered. Other factors affecting the amount of thermal energy delivered to users (and corresponding revenues) per dollar invested in the distribution system include:

1.   Intensity of land use as related to service requirements per unit of land area
2.   The degree of market penetration for community system services within its service area
3.   Peak demand and annual consumption of services
4.   The physical plan or layout of the community
5.   The location of the community system plant

Thus, the status and physical characteristics of a community are of primary importance in determining if it is economically feasible to distribute cogenerated thermal energy to the user buildings. In concept, considerations are essentially the same as for district heating. In practice, most TE applications have avoided many of the problems by serving new communities under central management. Whether owned by the land developer, a utility, or a third party, system planning has been integrated with community development from the start. Most exceptions are special community types: single buildings with a central HVAC system and medical, educational, or government complexes with existing district heating. With other community types, one of the first steps in system planning should be to examine the institutional and economic feasibility of distributing thermal energy.

B. Service Requirements

Total Energy and other community systems concepts would typically be located in a central equipment building, but some major components and subsystems could be distributed within the communities. A plant designer would typically draw some schematic boundaries around the plant and then determine the material and energy inputs and outputs that are required. Once these data are available, the designer can proceed with equipment selection and subsystem integration within those bounds. At this point, the community is characterized by the types of services required and the service demand profiles for each type.

Some community characteristics are related to the geographical location of the site. These factors include the type and availability of fuel supply, the availability and possible interconnections with conventional utility systems, and the availability of other resources such as water and land. Services provided by cogeneration would at least include electricity and heating and cooling. Depending on the locality, it may be beneficial to integrate the treatment and disposal of liquid and solid waste with the cogeneration plant.

Detailed community characterization, together with local weather data, again become important in determining service-requirement profiles. The plant designer needs the design peak demand and usage profiles for domestic electricity (excluding HVAC loads), the combination of space and water heating, space cooling, and solid and liquid waste production. Ideally, these data could be obtained from measurements from the community to be served or a similar community, but it is more likely that estimates would have to be calculated from known architectural characteristics of buildings, the intended use of buildings, estimated diversity factors which would account for the noncoincidence of load, and local weather data. More detail is usually required than for the design of conventional utility and building HVAC systems. A cogeneration plant produces heat coincidental with electric generation, but the demand of the community for heat is a function of weather and the occupancy of the buildings. Many plant designers and system analysts make use of computer-developed hourly demand profiles and system heat balances to properly account for the noncoincidence of demand and supply of thermal energy. The possibilities of using energy recovery from solid waste, thermal storage, various combinations of utility interties, and load management techniques within the community make hourly computer techniques more valuable.

Another type of load profile that must be considered is the change in service requirements over time as the community develops. A prime advantage of community systems is installation in phase with the development of the community, but some advance planning is necessary to properly select and size components and to schedule their installation.

III. TECHNOLOGY CONSIDERATIONS

Community characteristics are site-specific and difficult to generalize, but once determined, can be used with technical characteristics of equipment to make initial selections and a conceptual design.

A. Total Energy Design

Total Energy (TE) typically refers to a community cogeneration system that normally operates independent from conventional utilities. System operation follows electric demand, and thermal energy is generated as a by-product. The number and capacity of generator sets would be determined by the peak critical electric demand and the criteria to supply that peak demand with high reliability and without use of utility power. Additional considerations include unit sizes that are commercially available, minimum and average electric loads (so operating units run at high load factors), and expected community growth. High reliability requires the use of multiple components and excess capacity.[1] A system having at least two excess generator sets could still meet the critical load if there was a malfunction of one operating set while another was out of service for routine maintenance. The optimum level of spare capacity will be dictated by the inherent reliabilty of the equipment employed, the frequency and duration of peak loads, and the tolerance the system users have of outages. Some installations have operated successfully with only one spare unit. For localities with several installations serviced by one organization, it may be possible to use portable equipment as spares and reduce installed standby capacity.

The reliability of other services is typically provided by the use of excess auxiliary boiler capacity and multiple units for other major components.

The size of generator units selected for a TE plant and plant load characteristics have a significant bearing on the selection of generator prime movers. As an example, the applications of interest in the MIUS program were residential-commercial developments with loads equivalent to roughly 300 to 3000 dwelling units. For a garden-apartment model with a Philadelphia climate, the corresponding electric demand (including plant auxiliaries required for heating and cooling) ranged from about 750 to 7000 kW, and the size of individual generators ranged from about 250 to 1800 kW.[4]

Present economic and technological considerations limit the choice of prime movers in this size range to gas and steam turbines, spark-ignition gas engines, compression-ignition Diesel engines, and dual-fuel engines. The choice between these units will depend on many factors, such as the ratio of the heat-to-electrical demand and usage, fuel availability and costs, load factors, and the fixed charge rate on capital investment. The most important factor is the relative demand and use of heat and electricity and the required temperature of the heat. In the power range of interest, the internal-combusion engines are most suited for Total Energy installations for dwelling units where the primary product is electricity and the low-grade heat recovered is the by-product. Small steam turbines are of interest in applications that require large quantities of process heat, and the electrical output is a by-product (heat-to-electric use ratios of 15 to 20). Gas turbines are between these two extremes, and are of interest in installations in which the demand and usage of heat is about four to five times that of electricity. It is of interest to note that the *1971 Total Energy Directory and Data Book*[7] lists about 500 TE installations of various types in the U.S., using over 1400 prime movers. The percentages of the different types of prime movers installed in these systems are 70% gas engines, 15% gas turbines, 7% diesel engines, 7% dual-fuel engines, and 1.5% steam turbines. It should be emphasized that most of these plants are small, with outputs in the range of a few hundred to a few thousand kW of installed capacity.

A 1971 survey[8] of power-generating plants using internal combustion engines shows only 2% with gas engines, 64% with dual-fuel engines, and 34% with oil-fueled diesels. Most of these plants have installed capacities of 5000 to 50,000 kW, with engine-generator sizes in the range of 1000 to 6000 kW.

Although the primary attractive feature of TE is the potential for making use of cogenerated heat, there are technical and economic limits on the amount of heat that can be usefully recovered, and one must still be concerned with generating efficiency. A simplified comparison of fuel consumed in a TE subsystem of MIUS and a conventional system has been developed[9] to help judge the energy conservation merit of various TE prime movers. Figure 1 is a schematic of each system with efficiency assumptions indicated. As shown in the schematic, the comparison considers only the basic part of a TE plant, the generation of domestic and cooling-tower electric loads and the use of waste heat for heating and absorption chillers. Each set of the parameters N_e, R_i, and R_h corresponds to a combination of electricity and heating and cooling services delivered to the consumer per unit of fuel input (F). The amount of fuel required to deliver these same services with a conventional district heating and cooling system (shown schematically in Figure 1) can be calculated and compared with that for the TE model. The results, shown parametrically in Figure 2, illustrate the effect of MIUS generating efficiency and waste-heat utilization on the MIUS to conventional fuel consumption ratio. An analysis of Total Energy serving a hypothetical garden-apartment complex indicated that 52 and 58% of waste heat was used, and that 66 and 28% of that used was for air conditioning in Miami and Minneapolis, respectively. Results for Philadelphia and Dallas weather fell between these two extremes. According to Figure 2, the generating efficiency of the TE plant would have to be above about 20 to 25% in order to consume less fuel energy than a conventional district system. Considering the typical generating efficiencies listed in Table 1, only the internal combustion piston engines (and regenerative, compound, or combined gas-turbine cycles not listed) have adequate generating efficiency in small units. Simple-cycle gas turbines and the steam Rankine cycle are not attractive until unit capacity exceeds about 10 MW. Other community types, with different electric and thermal loads, and other conventional alternatives could, of course, lead to a different prime mover selection.

B. Other Operating Modes

Depending on the locality and particular application, Total Energy offers many advantages with respect to the simplicity of institutional arrangements and the close relationship between the community and the plant providing services. Unless electric and thermal loads are well matched, however, the maximum possible efficiency may not be achieved, and the excess capacity to insure reliability increases the capital cost. An intertie with the local electric utility is a logical extension of TE which offers several design options. Complete utilization of recoverable waste heat, and maximum use of fuel energy, is possible if the system operates to follow the thermal load, and electricity is considered a by-product. There are several possibilities, including selective energy, a two-way interchange, and use of cogeneration for utility peaking, but any case could probably be arranged to supply full emergency power to the cogeneration plant without expensive standby charges. If reliability is then supplied by the utility connection, then excess generating capacity would not be needed, and the designer could select one large generating unit instead of several smaller ones and, perhaps, a different prime mover.

There are many design options and economic trade-offs to be considered. The electric interchange rates during utility peak and off-peak hours would have a significant effect on the economically optimum mode of operation. In fact, the very existence of generating capacity in the community system may make it eligible to purchase low-

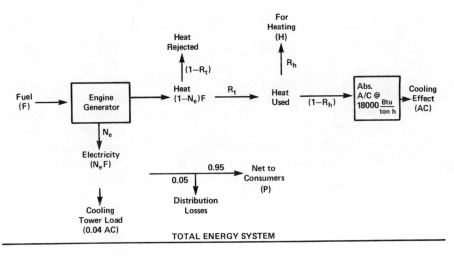

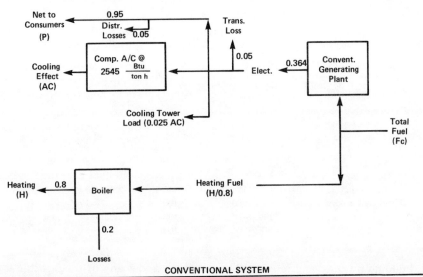

FIGURE 1. Energy flow schematic for simplified TE vs. conventional system comparisons.

cost off-peak power from conventional electric utilities. Because energy economics vary with time, are subject to sudden changes, and cannot be predicted with certainty, a strong case can be made to retain the flexibility of being able to operate in several modes. Thus, even with a utility intertie, a designer may want to use multiple units and provide space for additional capacity so that future independent operation would be possible.

IV. PLANNING AND OPERATION

A. Ownership Options

The flexibility that has thus far been claimed for community systems also applies to ownership and arrangements for operation and maintenance. In the past, Total Energy plants have typically been owned in common with the property served or by a private-sector entity established for that purpose. Potential owners include land developers,

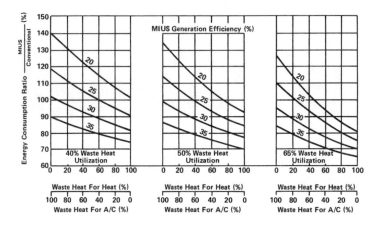

FIGURE 2.   Comparison of fuel consumption by MIUS with that of a conventional district system providing equal services.[9]

TABLE 1

Typical Generating Efficiencies of Small Units

Type	Capacity (kW)	Load (%)	Efficiency (%)
Internal combustion piston engines			
Gas-fueled	500	70	29.4
Diesel	500	70	32.2
Dual fuel	500	70	31.0
Gas turbine			
Simple cycle	500	70	17.0
Steam Rankine cycle	<1000	100	7—19
	1000—10,000	100	17—24
	10,000—50,000	100	22—31

builders, building owners, and tenants. These may be under any of the possible legal arrangements, such as wholly owned subsidiaries, partnerships, private corporations, or cooperative ventures. It would be unlikely for electric utility companies to be involved in the ownership of independent TE systems, but gas and district heating utilities are candidates.

With increasing interest in cogeneration as an energy saving concept and proposed incentives to encourage its wide application comes an increasing potential for utility company involvement in all types of applications. This could include utility ownership and operation of all or part of a community system, which is similar to the options for industrial applications.

The optimum ownership arrangement for any particular application will be decided by the local institutional characteristics, financing options, the economic requirements of each party, and the benefits and risks of the application perceived by each party. In any case, the successful application of community systems requires careful integration of planning to best phase system installation with community development. New community applications could require the use of interim services until loads are large enough to justify the installation of the cogeneration plant. One well-documented example[10] was a proposed HUD-sponsored MIUS demonstration to serve part of the new town of St. Charles, Md. Several options for interim service were considered, but the preferred scheme was to initially install the central equipment building, the auxiliary boilers and electric-driven chillers, and the hot- and chilled-water distribution system to provide hydronic heating and cooling with purchased fuel and electricity (a conventional district system). Engine-generator sets with waste-heat recovery would then be installed as the thermal demand exceeded the capacity of auxiliary equipment. This plant was possible because of intertie agreements with the local electric distributor. Other arrangements would be required for a stand-alone TE system.

B. Institutional Requirements

An important aspect of the HUD-sponsored MIUS Program has been the analysis of legal and institutional issues that would affect the implementation of MIUS. One approach[4] was to examine existing utility law as a basis for judging the legal treatment of a MIUS, since MIUS constitutes the integration of services with which the law is familiar and because several decisions on Total Energy systems are available for study. Furthermore, the flexibility inherent in the judicial process will most likely allow the expansion of existing principles to cover MIUS.

The great diversity among states in statute law, common law, legal precedence, and institutional relationships demands that each MIUS installation be considered separately. The need to treat MIUS on a case-by-case basis is stressed. However, a generic assessment was attempted by the use of the flow chart in Figure 3. The flow chart shows a generalized set of legal issues faced by the developer of a MIUS. By considering each of the issues presented, the developer should be able to determine if possible exclusionary provisions of the law, or the burden of future regulation, would preclude his implementation of MIUS.

As indicated, a developer must first determine the proper legal entity for his proposed MIUS. It was concluded that existing forms, such as partnerships or corporations, are suitable, and that there should be little difficulty in acquiring a legally recognized status.

Having resolved this, the developer is faced with the question of whether a MIUS would be considered a public utility. A range of legal obligations, liabilities, and outside influence stems from this classification. The most important is centered on regulation, taxation, and condemnation. For this assessment, public utilities are considered to be those franchised and regulated by state public service commissions. Municipal utilities are considered to be municipally owned or operated utilities that, although public in nature, are not directly regulated by state agencies. Systems operated by nonmunicipal entities, which are not subject to direct regulation by state public service commissions, are considered private, or nonpublic, utilities. Whether such entities as private corporations or associations are classified nonpublic and free from the regulation of commissions depends on tests applied by the courts.

The approach to classification varies from state to state between two limits. In some states, providing a utility service to any portion of the public for compensation is sufficient for classification as a public utility. Other states determine utility classification by type of ownership. Exemption of cogeneration plants from federal and state public utility regulations is a proposed incentive for the concept.

As a public utility, a MIUS would need a franchise from the state in which it is located, to be cognizant of existing utility territories that cannot be violated, and to be subject to state regulation. Benefits pursuant to a public utility classification include protection from competition, certainty in established regulatory procedures, and the power of eminent domain.

Certain competition-related factors are so restrictive that operation of a MIUS is precluded in some areas. The possibility of MIUS exclusion is represented by blocks 3 and 4 of Figure 3.

Finally, the nature of regulation which might be experienced by the MIUS developer

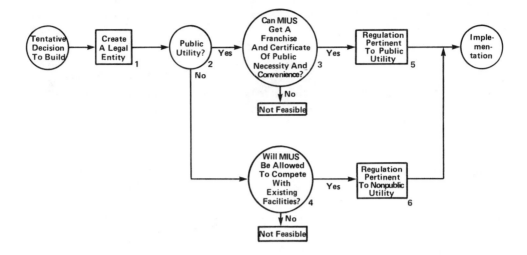

FIGURE 3.   Legal decisions required for the implementation of a MIUS system.

is shown in blocks 5 and 6 of Figure 3. The public utility classification carries with it a greater burden of regulation that directly affects the creation and eventual operation of the MIUS. However, there is no certainty that, having avoided the public utility classification, the MIUS can totally avoid similar regulation.

In conclusion, it was apparent that legal constraints would not preclude a MIUS. On the other hand, a MIUS will not receive uniform treatment by the law due to local variations of law and procedure. To accommodate local variation, a developer will have to exercise caution in organizing the MIUS. It appears that at least a part of the potential market for MIUS will be eliminated, either by exclusion or by the unwilling-ness of a developer to assume the burden of regulation. Such a burden may be better shouldered by a conventional utility serving a large public than by a small MIUS with a limited staff. The effects should be minimal, however, if MIUS is owned by conven-tional utilities, municipalities, or large private concerns (perhaps operating several MIUS installations) that have the experience and the staff to deal with legal and regu-latory requirements.

A second approach of the MIUS program was to include an analysis and documen-tation of institutional issues and their resolution for a site-specific installation, a pro-posed MIUS demonstration at St. Charles, Md.[11,12] A principal objective of the grant, which carried the project through preliminary design, was to prepare an example for obtaining institutional approvals which could be used by others. This site-specific case had the advantage of not being a hypothetical study. The intended owner, Interstate Land Development Company, Inc. (ILD), aggressively pursued the approvals needed for MIUS with the intent of building and operating the system.

The proposed MIUS demonstration at St. Charles was to provide electricity, space and water heating, space cooling, solid-waste incineration with heat recovery, liquid-waste treatment, and irrigation of a local golf course. The community was a moderate-density 130-acre village center consisting initially of 417 dwelling units, a 900-student middle school, a 100,000 ft² village center of retail, office, and community facilities, and a 20,000 ft² MIUS building. Electricity produced by the plant would be used for internal plant loads and the village center. Excess would flow to the local electric dis-tributor, the Southern Maryland Electric Cooperative (SMECO). SMECO would pro-vide electric service to residential units and full standby service for the MIUS.

ILD classified each institutional interface according to its influence on approval, design, and economics for each service. A summary of results is shown in Table 2. It

TABLE 2

Potential Influence of Institutional Factors on a St. Charles MIUS

Institutions	Electrical A	D	E	Thermal A	D	E	Waste water A	D	E	Solid waste A	D	E
Federal												
HUD (NCA & FHA)				x	x	x		x				
U.S. EPA							x		x			
Rural Electrification Administration (REA)	x		x									
National Rural Utilities Co-operative Finance Corp.	x		x									
U.S. Dept. of Energy (DOE)	x	x	x		x	x					x	x
U.S. Internal Revenue Service (IRS)		x				x			x			x
State of Maryland												
Maryland Public Service Commission (PSC)	x	x	x				x	x	x			
Dept. of Health & Mental Hygiene		x	x				x	x	x	x	x	x
Dept. of Natural Resources (DNR)	x						x	x	x		x	x
Div. of Labor & Industry (OSHA)		x	x	x	x			x	x		x	x
Dept. of Assessment & Taxation		x				x			x			x
Maryland Dept. of Planning												
Maryland Energy Policy Office												
Maryland Board of Public Works												
Maryland Board of Certification							x		x			
Charles County Govt.												
County Commissioners	x		x	x		x	x		x	x		x
Planning Commission	x	x	x	x	x	x	x	x	x	x	x	x
Dept. of Parks & Recreation								x	x			
Dept. of Environmental Health							x	x	x	x	x	x
Dept. of Public Works (PWD)							x	x	x		x	x
Board of Education						x					x	x
Misc. County Inspectors and Authorities	x			x			x			x		
Misc. organizations												
Southern Maryland Electric Co-op (SMECO)	x	x	x									
Potomac Electric Power Co. (PEPCO)	x		x									
St. Charles Design Review Committee (DRC)	x	x	x	x	x	x	x	x	x	x	x	x
Aetna Insurance Co.		x	x		x	x		x	x		x	x

Note: A, approval influence; D, design influence; and E, economic influence.

From Reeves, W., Preliminary Design Institutional Analysis Report—MIUS Demonstration, St. Charles, Maryland, Interstate Land Development Co., Inc., for the Department of Housing and Urban Development, HUD-PDR-373-2, September 1978.

was concluded that a MIUS (or other community system) project would typically start by establishing institutional feasibility and support at the local government level. In St. Charles, this entailed working with the County Planning Commission and the St. Charles Design Review Committee, both of whom had considerable influence because of their authority over the type and intensity of land use and the plants' architectural treatment.

It was also considered essential to gain early support of the local electric utility and then to apply to state-level agencies for needed approvals. With the cooperation and assistance of SMECO, a mutually beneficial intertie arrangement was formulated by which the MIUS was ruled to be exempt from Maryland Public Service Commission regulation. This was considered a major breakthrough because it freed the project from the submission of an environmental impact study for consideration by the Power Plant Siting Committee.

Although institutional issues affected MIUS design and economics, it appeared that all such issues could be resolved in ways to move the project past potential barriers. However, due to budgeting limitations for the program in HUD, no further work beyond the study documentation is planned.

REFERENCES

1. Samuels, G. and Meador, J. T., MIUS Technology Evaluation—Prime Movers, ORNL/HUD/MIUS-11, Oak Ridge National Laboratory, Oak Ridge, Tenn., April 1974.
2. Segaser, C. L., Internal Combustion Piston Engines, ANL/CES/TE 77-1, Oak Ridge National Laboratory for Argonne National Laboratory, Argonne, Ill., July 1977.
3. Segaser, C. L., Heat Recovery Equipment for Engines, ANL/CES/TE-77-4, Oak Ridge National Laboratory for Argonne National Laboratory, Argonne, Ill., April 1977.
4. Mixon, W. R., Ahmed, S. B., Boegly, W. J., Jr., Brown, W. H., Christian, J. E., Compere, A. L., Gant, R. E., Griffith, W. L., Haynes, V. O., Kolb, J. O., Meador, J. T., Miller, A. J., Phillips K. E., Samuels, G., Segaser, C. L., Sundstrom, E. D., and Wilson J. V., Technology Assessment of Modular Integrated Utility Systems, Vol. 1, Summary Rep. ORNL/HUD/MIUS-24, Oak Ridge National Laboratory, Oak Ridge, Tenn., December 1976.
5. Meador, J. T., MIUS Technology Evaluation—Comparison of Trenching Costs for Separate and Common Trenching, ORNL/HUD/MIUS-15, Oak Ridge National Laboratory, Oak Ridge, Tenn., February 1976.
6. Tison, R. R., Blazek, C. F., and Biederman, N. P., Advanced Coal-Using Community Systems, Task 1A. Technology Characteristics, Vol. 2, Institute of Gas Technology, Chicago, Ill., March 1979.
7. Total/Energy, 1971 Director and Data Book, *Vol. 8(1), Total/Energy Publishing,* San Antonio, Texas, 1971.
8. 1971 Report on Diesel and Gas Engines Power Cost, *American Society of Mechanical Engineers,* New York, 1971.
9. Payne, H. R., MIUS Technology Evaluation—Lithium Bromide-Water Absorption Refrigeration, ORNL/HUD/MIUS-7, Oak Ridge National Laboratory, Oak Ridge, Tenn., February 1974.
10. Reeves, W., Preliminary Design Technical Report - MIUS Demonstration, St. Charles, Maryland, Vol. I, Book 1 of 2, Interstate Land Development Co., Inc., St. Charles, Md., for the Department of Housing and Urban Development, HUD-PDR-373-1, September 1978.
11. Rothenberg, J. H., A Review of the HUD Total Energy experience and the MIUS program, 144th Annu. Meet. Am. Assoc. Advancement Sci., Washington, D.C., February 13, 1978.
12. Reeves, W., Preliminary Design Institutional Analysis Report for a HUD/MIUS Demonstration Project, St. Charles, Maryland, Vol. II, Book 1 of 1, Interstate Land Development Co., Inc., St. Charles, Md., for the Department of Housing and Urban Development, HUD-PDR-373-2, September 1978.

National Implementation Considerations

SECTION 4

NATIONAL IMPLEMENTATION CONSIDERATIONS

PREFACE

Cogeneration systems for either industrial or district heating applications are obviously desirable from energy and environmental points-of-view. It is apparent, however, that both technical and institutional problems will limit the implementation of this technology in many applications.

Among the problems associated with cogeneration implementation are questions concerning the regulatory and legal aspects of any proposed cogeneration facility. In spite of the fact that industrial and district heating cogeneration has been practiced in this country for many years, these questions are not all obvious, nor are their solutions apparent. Itemization and discussion of many of these problems are given in Chapter 13 along with suggestions for strategies which might be useful in resolving such problems .

The siting of cogeneration plants is obviously a question which involves the balancing of energy costs vs. transmission costs for both heat and electric power. A brief analysis of this question is given in Chapter 14, along with an example of methodology useful for the identification of sites where there exists sufficient energy demand within reasonable distances.

The degree to which cogeneration can be incorporated into our society has been studied by several workers with varying degrees of optimism. One such study considers the expansion of industrial energy systems and evaluates the cogeneration potential for industrial uses as a function of various implementation strategies. A summary of this study is given in Chapter 15.

Similarly, the potential applicability of district heating and Total Energy systems may be evaluated in terms of existing locations and projected growth rate. An analysis of district heating potential is presented in Chapter 16. The potential for total energy systems is further analyzed and projected as a function for housing starts and fuel costs in this same chapter.

Chapter 13

LEGAL AND REGULATORY CONSIDERATIONS*

Norman L. Dean, Jr.

TABLE OF CONTENTS

* This chapter is excerpted from *Energy Efficiency in Industry* to be published soon by the Ballinger Publishing Co. of Cambridge, Mass. The book was prepared under a grant from the National Science Foundation and is copyrighted by the Environmental Law Institute. All rights reserved.

I. INTRODUCTION

Companies desiring to cogenerate electricity and useful heat have met numerous legal and regulatory barriers. Most of these barriers relate to the need of the cogenerator to receive backup utility service or to dispose of the electricity it generates in excess of its own on-site needs. There are four major types of these utility law barriers, (1) the monopoly status granted to incumbent utilities, (2) the lack of incentives for cooperation by existing utilities, (3) the level of utility rates, and (4) the possibility of government regulation of private companies as "public utilities." This chapter briefly outlines each of these major types of barriers and discusses policy initiatives currently under discussion or enacted for lowering these barriers and encouraging cogeneration. Of course, legal and regulatory barriers constitute only one of the types of institutional barriers that have impeded cogeneration. Attitudinal and economic barriers have also stood in the way of cogeneration. These barriers are treated in other chapters of this book as well as in other recent publications.[1-6]

II. THE MONOPOLY STATUS OF EXISTING UTILITIES

Utilities, unlike other industries, have been subject to government regulation under the theory that they are "natural monopolies". The states have attempted to protect consumers from the consequences of this natural monopoly by expressly granting utilities monopoly status in exchange for controlling their rates and operation. The monopoly status granted to existing utilities generally protects those utilities from retail competition by other "public utilities". This monopoly status is clearly one of the most serious legal barriers to widespread industrial cogeneration. If industrial cogeneration plants that attempt to sell power are defined by state law as "public utilities", then they may be prohibited from competing with the existing utilities.

A. Statutory Protection of Monopoly

In a few states, the monopoly status of existing utilities is enshrined in the state statute books. In such states, the state public utilities commission (PUC) may authorize a utility to serve an area already being served by another utility only if the existing utility's service is found after a hearing to be both inadequate and unlikely to be made adequate.[7]

B. Protection of Monopoly Through Certification

In most states, however, the monopoly status of existing utilities is not mandated by statute, but rather, has been developed by the state PUC in fulfillment of its responsibility to regulate in the public interest. To prevent the wasteful duplication of facilities, to prevent diminished efficiency and high costs, and to insure reliability, the PUCs have adopted the "regulated monopoly" concept. The tool used to implement this policy is typically the "certificate of public convenience and necessity."

In many states a public utility must receive a certificate of public convenience and necessity before it can begin the construction of a new facility, or before it can begin construction of its first utility plant. In most of these states, the PUC has substantial discretion in deciding how and on what basis to issue a certificate. Typically, certification proceedings inquire into the need for the facility and the ability of the applicant to provide safe, adequate, and reliable service. They examine the technical and economic viability of the proposed project. However, the decisive question is often whether an incumbent utility is capable of serving the area. If there is such an incumbent utility, application for a certificate will usually be denied.

If an industrial cogeneration plant is held to be a public utility, then it would be forced to apply for a certificate. Not only would such an application be time consuming and expensive, but the existing utilities may oppose such certification, and in general, the PUCs can be expected to give preferential treatment to the existing utilities. This is because of the widely held rule that " . . . a utility in the field shall have the field unless public convenience and necessity require an additional utility."[8] Even where the incumbent utility is providing deficient service or no service at all it will be given preference in a hearing on public convenience and necessity. If an existing utility is providing inadequate service many PUCs will give it a period of grace in which to remedy its deficiencies. And where an applicant for a certificate seeks to serve an area not presently being served by a utility many commissions will give preference to a nearby existing utility.[9]

C. Protection of Monopoly Through Franchise Issuance

Another tool widely used to protect the monopoly status of incumbent utilities is the franchise. The franchise is a right granted by a state or local government to a corporation or individual permitting it to use or occupy the streets or other public property. In many, if not most, jurisdictions, a company will not be permitted to generate and distribute power without a franchise if that company's lines use, cross, tunnel under, or in any way encroach upon public property, streets, or alleys.[10]

Many persons confuse the franchise with the certificate of public convenience and necessity (or the territorial allocation) issued by state public utility commissions, but the two are decidedly different. The franchise is usually issued by a local government under authority derived from the state legislature or constitution, although it is sometimes issued by a state government. It is often seen as a valuable property right that cannot be taken without compensation. The utility commission certificate, on the other hand, is usually seen as a mere license that can be taken away. It is not issued by a state or local government, but rather, the utility commission. In many states, a public utility must have both a certificate of public convenience and necessity and a franchise before it can supply customers.

In accepting a franchise, a company must often dedicate itself to serve an area adequately. In some cases, franchised utilities have been required to expand their service to meet the needs of growing customers. Local governments can and sometimes do employ their right to refuse to issue a franchise to protect the monopoly of existing utilities.[12]

Typically, local governments are given extremely broad latitude in determining whether or not to issue a franchise. The broad discretion that has been given to local governments by statutes such as these makes it difficult to generalize about their likely impact on cogeneration projects. There appear to be several different ways in which the requirement for a franchise can adversely affect cogeneration projects.

First, it may be difficult, time consuming, and expensive for a private industry to obtain the franchise. Franchises are often issued by the vote of the local legislature, and sometimes require a referendum of the voters.[13] This political process can conceivably cost a great deal of time and money.

Second, an industry may be unwilling to subject itself to the restrictions and conditions that a local government may demand.

The third way in which the franchise requirement can adversely impact a cogeneration project is by blocking it altogether. The local legislature may for one of a thousand political reasons simply refuse to issue the franchise, or the legislature may be powerless to issue one where an incumbent utility holds an exclusive franchise. In some states such as Oklahoma, the statutes, constitution, or case law prohibit the issuance of an exclusive franchise.[14] However, in roughly one half of the states, local governments can issue exclusive franchises.[10] Where an incumbent utility holds an exclusive franchise, the local government will likely not be able to issue an industrial cogenerator a franchise without either receiving the permission of the incumbent or exercising its power of eminent domain and compensating the incumbent.

Even when the incumbent utility holds a nonexclusive franchise, it can sometimes employ the franchising process to protect its monopoly and block a cogeneration scheme. This danger is illustrated by a recent Oklahoma Supreme Court case.[11]

Does industry that has been denied a franchise or that objects to the terms of a franchise have any recourse? Some states provide for an appeal by the applicant for a franchise where that applicant feels it has been treated unfairly by a local government. Wisconsin law, for example, provides that an applicant may appeal to the state public service commission (PSC).[15] If the PSC finds that the franchise is unreasonable, the PSC can declare it to be void. However, in many states, an applicant that has been denied a franchise by a local government cannot receive one from the PUC.

III. UTILITY RATE STRUCTURES

While the regulated monopoly concept has prevented industry from selling cogenerated power at retail, utility rate structures have worked against the industrial cogeneration of power generally — whether for use within the cogenerating plant, for sale at retail, or for sale to utilities at wholesale.

An industry's decision as to whether to build a cogeneration facility is extremely sensitive to the existing level of utility rates. In particular, the industry decision to invest or not to invest in a cogeneration plant will be greatly influenced by:

- The price the utility is willing to pay for excess electric power generated by the industry
- The rates the utility charges to the industry for standby power that the industry needs during those periods when its cogeneration equipment is being repaired or maintained
- The rates the utility charges for normal service to industrial customers.

Obviously, cheaper regular industrial power rates mean reduced savings to the industry that generates its own electric power.

In addition, the industry's view of the profitability of a cogeneration investment might be affected by the rates charged for the wheeling of electric power and the rates an industry is permitted to charge any third parties to which it sells electricity (i.e., another industry, a shopping center, a university).

The reform of electric utility rates is far too complex a subject to be dealt with in-depth in this chapter. This section simply hopes to make the point that utility rates have a significant impact on the implementation of industrial cogeneration, and that there is a serious question as to whether those rates have been fair to industrial cogenerators. At the least, there is a need for utility commissions to examine utility rates to insure that they do not discriminate against industrial cogeneration facilities, and that they are set on the basis of the utility's costs. While several PUCs have undertaken extensive reviews of the utility rates charged to customers, few have examined the rates at which standby power is charged or the rates at which utilities are willing to purchase privately generated power.

A. Rates for Normal Industrial Service

An industry's decision to cogenerate or not to cogenerate will plainly be affected by the price of the alternative source of supply, which is typically power purchased from the public utility. Utility rates are generally based on the average cost to the utility of generating power rather than the replacement cost (the cost to the utility of producing power from a new, relatively expensive generating plant). These average costs provide price signals to industry that undervalue cogeneration. An industry may be asked to pay 3¢/kWh for its power when the utility must pay over 4¢/kWh to generate power from its newest generating facilities. An industry that can cogenerate for 3.5¢/kWh may thus choose to purchase the cheaper utility-generated power despite the fact that it would be "cheaper" for the nation if the industry generated its own electricity.

B. Rates for Standby Power

Whether generating power for their own use or for sale to others, most industries will want to remain connected to the utility so as to be able to purchase emergency power in the event that their own equipment fails or is shut down for maintenance. Such standby power connections have been discouraged by high demand charges, ratchet rates, and special standby rates that make standby power many times higher than regular utility rates.

Most large industrial customers are billed for electric power under a two-part rate scheme that includes a demand charge and an energy charge. The demand charge is a fee for the maximum amount of power that the industry used during the billing period, usually measured as the highest level of consumption during any 5, 15, or 30 min interval during the billing period. It is intended to compensate the utility for the fixed costs of supplying power, such as the equipment that it must keep ready to serve the customer. In addition, the industry is assessed an energy charge, a fee for each kWh of electricity that the industry consumed during the month. It is intended to compensate the utility for the variable costs of providing power, such as fuel. Many utilities charge a rather large minimum demand charge whether any power is actually consumed during a month or not. Moreover, some rate structures for large industrial customers include a so-called "ratchet rate" that discourages the use of relatively high amounts of power for very short periods of time by carrying over the demand charge into future months. Under the ratcheting rate used in California, for example, an industry is assessed a charge set at 50% of the highest maximum demand established during the previous 11 months.[16]

Demand charges (including the ratchet rate) require industries that use a large quantity of electricity for a short period of time to pay much higher rates. The Maine cogeneration study, for example, found that a cogenerating paper plant that was required to purchase all of its demand for short periods of time would have to pay 6.3¢/kWh instead of the "normal" industrial rate of 2.25¢/kWh.[17] The New Jersey study found that 1.5 MW of backup power for a cogeneration unit in a rubber plant operation would be about 8¢/kWh instead of the 3.4¢/kWh rate that the industry would have been charged had it purchased all of its power needs from the utility.[5]

In some states, industrial cogeneration plants that use power only occasionally for backup or standby service will also be subjected to a special standby assessment. The standby assessment is for utility breakdown service where the entire electrical requirements on the customer's premises are not regularly provided by the utility. Standby assessments are assessed in addition to the regular demand charge and energy charge, whether or not any electric power is consumed by the industry. Standby assessments can be extremely high. Southern California Edison, for example, charges $2.00/month for the first 20 kW of power that the customer may need as standby and $1.50 for every kW in excess of the first 20.[18] Pacific Gas and Electric charges $2.74/kW for the first 25 kW, $2.08/kW for the next 100 kW, and $1.64/kW for anything over 125 kW. San Diego Gas and Electric charges a total of $68.58 for the first 20 kW and $2.74 for each kW over the first 20.[16]

Standby, demand, and ratchet charges are necessary to compensate a utility for the cost of the generating equipment that it must keep ready for that moment when an industry suddenly demands its standby power. However, demand charges and standby rates are generally not based on the cost to the utility of providing standby, since the cost to a utility of providing backup power varies depending on the time of the day and of the year that the industry demands that power. Present demand and standby rates typically do not vary based on the time of day or year. If an industry demands standby at the same time most of the utility's other customers are demanding power, then it should be assessed relatively high demand charges. However, if the industry demands standby power at a time when the utility has a significant amount of generating equipment sitting idle, then the industry should be assessed little or no standby charge.

There is a need for PUCs to examine demand charges and standby assessments that are levied against cogeneration facilities. Those charges should be set so as to reflect the actual cost to the utility of providing the standby service.

In determining these costs two points should be kept in mind:

- Standby rates should be based on probabilistic analysis of when and where cogeneration facilities will be shut down. The rates should not be calculated based on the assumption that all cogeneration facilities in an area will shut down simultaneously. Lower rates will doubtless be charged for standby power as a greater number of cogeneration units begin operation in a utility's service area and several industries are able to share the capital costs of standby equipment.
- It follows, then, that standby rates should be based on time-of-use pricing principles. Charges should vary with the time of day and the season of the year, since cogeneration standby service will require a utility to keep additional capacity ready only when the use of that standby service corresponds with a utility's peak demand period.[19]

The state commissions should make every effort to develop innovative billing techniques for giving industry the ability to reduce their standby charges. For example, industries could be offered "interruptible standby" rates. With interruptible rates, the utility would promise to use best efforts to provide an industry with standby power if the utility had the available fuel and equipment, but if not, the industry would have to do without power. PUCs should also examine the possibility of providing lower standby rates to those industrial customers who are willing to accept "scheduled standby" or standby that would be made available only during certain times of the month, day, or year. This would enable industries to schedule the maintenance of their cogeneration equipment during the least costly time.

C. Rates for Excess Power

As noted earlier in this book, gas turbine or Diesel cogeneration systems may produce significantly more electricity than industries can consume on-site. Utilities have shown a general lack of willingness to pay reasonable rates for excess power generated by industry. For example, the Stauffer Chemical Company approached the Pacific Gas and Electric Company (PG&E) with an offer to sell power that it was producing using waste heat from a sulfuric acid plant. PG&E offered first to buy the power at

0.2¢/kWh. After some negotiations, it offered 0.3¢/kWh, and finally offered 1.4¢/kWh. This compared with PG&E's average generating costs of about 3.0 to 3.6¢/kWh for oil-fired generators and 2.4¢/kWh for nuclear.[20]

The guiding principle for the price of utility-purchased power should be that the utility should offer to purchase "firm power" (power that will be available under a long term contract at times when it is needed by the utility) at a price equal to the costs that the utility would incur by generating additional amounts of electricity at that time of day and year. If the use of cogenerated power saves the utility from having to build additional power plants and other capital equipment, then the rates paid to the industry should be appropriately higher. Overall, the rates for purchased power should be set so that they encourage cogeneration when it is the lowest-cost alternative source of electric power supply for the utility system as a whole.

Ideally, state PUCs should require that utilities use these guiding principles to devise rate schedules that state the prices at which utilities are willing to purchase industrially generated power for each time of the day and year. It is likely, however, that many PUCs will feel uncomfortable regulating the rates that utilities offer for industrially generated power. PUCs have traditionally left most day to day operational decisions to utility management. For example, the PUCs have rarely involved themselves in dictating the rates at which the utilities can purchase fuel or generating equipment. Also, few statutes explicitly grant PUCs direct authority over the rates of purchased power. However, most commissions clearly have authority to *indirectly* control the rates of purchased power. They can, for example, deny the utility the right to include the cost of fuel purchases and other expenses in its cost of service for rate-making purposes when the utility has failed to offer to purchase power at reasonable rates. Also, the PUC can refuse to permit the utility to build new generating plants.

The California Public Utilities Commission is apparently the first to encourage development of rates for industrially cogenerated power. In January 1978, the California PUC ordered the three major California utilities to submit "guidelines covering the price and conditions for the purchase of energy and capacity from cogeneration facilities owned by others."[21] PG&E and SCE have responded with very general outlines of their cogeneration policies.

In its response, the Pacific Gas and Electric Company discussed the methodology it follows in arriving at a price for industrially generated power. The company divides industrially generated power into two major types. Dump energy is defined as that power which is available to PG&E at the seller's option, such as power available at random, during off-peak hours, and power that is interruptible by the seller or that cannot be scheduled. For dump energy, the company stated it will pay "up to its system-average energy cost less a percentage amount to cover other PG&E costs and to flow through a portion of the benefits to the customers."[22]

Reliable energy is defined by PG&E as energy that is largely available during the utility's peak or that is schedulable to some degree, that is supported by an assured fuel supply, that is available in defined quantities and rates of delivery, that is from a generating source whose maintenance schedule can be coordinated, that is sold under a long term contract, and that PG&E can refuse to purchase on 12 hours notice when less costly energy is available from other sources. The company also indicated a willingness to pay for reliable energy the "cost of energy from alternative sources less a percentage amount to cover other PG&E costs and to flow through a portion of the benefits to customers." To the extent that a source is able to provide something more than dump energy, but less than reliable energy, the company will adjust its price formulas accordingly.[22]

PG&E also expressed a willingness to pay a cogenerating industry for the equipment the industry uses in its cogeneration facility. Presumably, the value of this equipment to the utility is the value of the generating and other equipment that it will not have to construct as the result of purchasing rather than generating its own power. PG&E stated that for such industrial capacity to have full value it must possess the following characteristics:

1. There must be a defined amount of capacity to be purchased.
2. The term of the sale contract must be long enough for PG&E to defer building an alternative resource. Alternatively, a resource could have a near-term value if there is a projected shortfall in capacity available for meeting total system capacity requirements.
3. The capacity must be available during on-peak hours, with adequate energy to support it, and the resource must be fully schedulable by PG&E's dispatcher.
4. The resource must meet standards of reliability which conform to good utility practice as applied in the PG&E service area.
5. An adequate long-term fuel supply must be available.
6. In the event seller is also providing all or part of its own electrical requirements from its resource and it has a partial forced outage, the seller must curtail its own power usage to maintain availability of agreed upon capacity to PG&E.[22]

IV. UTILITY INCENTIVES

A number of commentators have suggested that the present system of utility regulation may discourage utility involvement in privately owned cogeneration projects. They have argued that since a utility cannot include privately owned equipment in its rate base, it will have little incentive to join in such efforts. There may be other reasons for utilities to ignore private cogeneration projects, including the existence of large reserves of excess capacity or fear of loss of status on the part of utility management. On the other hand, utilities may see in cogeneration an opportunity to overcome financial and power plant siting problems and to respond more rapidly to unexpected changes in demand. It is difficult to sort out these competing incentives and predict how utilities will react to cogeneration over the long term. If history is the guide, utilities can be expected to eye cogeneration projects suspiciously, and policy makers may be forced to devise tools for encouraging their cooperation. A number of such tools have already been proposed, ranging from laws that would permit utilities to include privately owned equipment in their rate base, to tax incentives, to laws that would permit or require PUCs to mandate utility cooperation.

V. INDUSTRY FEAR OF REGULATION AS A PUBLIC UTILITY

Many industrial executives have stated that they are unwilling to build cogeneration plants out of fear that by generating electric power they will become "public utilities" under state or federal law and thereby subject to regulation by the PUC or FERC.

A. Consequences of PUC Jurisdiction

What exactly is it that these industrial executives have to fear? The list of possible consequences of state PUC jurisdiction is long, including the following.

1. The Need for a Certificate of Public Convenience and Necessity

As discussed above, in many states a public utility must receive a certificate of public convenience and necessity before it may begin construction of a new generating facility or construct its first plant.

Certification proceedings often inquire into the safety, reliability, and adequacy of proposed facilities. For a cogeneration plant, the proceedings would likely center on the reliability of the power to be produced. PUCs will be interested in such questions as:

- Does the industry have a secure supply of fuel? In the event of a nationwide fuel shortage, will the industry supply be cut off by fuel allocations?
- How secure is the industry's demand for its primary product? If demand drops, will the industry shut down its process lines and with them the cogeneration unit?
- Are labor disputes likely to close down the industrial cogeneration facilities?
- How technically reliable is the equipment to be used by the industrial cogenerator?
- Can the electrical output of the plant be adjusted to meet changing consumer demand?

If the cogeneration facility's reliability is considerably less than central station utility power, then many PUCs will be understandably reluctant to grant certification. Fortunately, it appears that with careful planning most cogeneration facilities in private industrial facilities can be made highly reliable.

2. Rate Regulation

If declared to be a public utility, the industrial firm selling power would be required to have its retail rates approved by the PUC and may have to have its wholesale rates approved by FERC. Industrial firms fear that they would not be permitted to increase their rates fast enough to keep pace with increasing costs of fuel, equipment, etc. Moreover, there is uncertainty as to how the "just and reasonable standard" of rate setting which has been developed to apply to regulated public utilities would be applied to industrial firms generating power. Would the rate of return to an industrial cogeneration facility be set commensurate with the returns being earned by utilities generating electric power, or commensurate with the returns being earned by other industrial facilities? This is a noteworthy question since utilities typically earn lower returns than industries.

3. Duty to Serve

State laws universally require a public utility to serve all customers in its service area that demand service. These statutory provisions raise the possibility that an industrial cogeneration plant will be required by law to expand to meet the new or increased needs of its customers, whether or not the industry has the requisite waste heat or process steam needs, or the desire to produce additional power.

4. Inability to Discontinue Service

Most states require that once a public utility has begun providing electric power to the public, it may not stop providing utility service without the express permission of the utility commission. Conceivably, an industry that signed a five-year contract to provide electric power to a housing development or other industry could be ordered to continue providing the power even after the expiration of the contract.

5. Restriction on the Sale of Securities

In many states, public utilities may not issue or sell securities without prior approval of the PUC. It is uncertain how these provisions of law could or would be applied to industrial cogeneration plants.

If an agreement between a business and utility were found to be a security, the PUC would look to the amount of the debt and the purpose of the issuance in determining whether to approve their sale or offering. Since some states require that the issuance of securities by a utility be by competitive bidding, the utility conceivably might have to hold out to public bid its offer to enter into a long term power contract for the purchase of industrially generated electric power.

6. Mandatory Reports and Accounts

Utilities are universally required to file various reports, records, and accounts with the PUC. Materials filed by an industrial cogeneration plant might contain proprietary materials that could harm the industry if they fell into the hands of a competitor.

7. Emergency Controls

Certain states provide the PUC with selected emergency powers, such as the power to require a utility to interconnect with and sell power to another utility or to a private customer. Several industrial companies have expressed the fear that they could be forced to cut back on manufacturing, increase their production of power, and sell to residential or other customers in the event of a future severe power shortage.

8. Construction and Equipment Standards

Most public utility commissions require regulated utilities to comply with certain minimum safety standards. Many also require approval of the safety and adequacy of construction plans.

VI. THE UTILITY STATUS OF COGENERATION FACILITIES

Unfortunately, it is not easy to state under what conditions an industrial cogeneration plant will become subject to the jurisdiction of the state PUC or FERC. Regulatory jurisdiction hinges on statutes that are usually very general in their language, sometimes ambiguous, and often subject to varying interpretations. Jurisdiction may vary depending on the location of the plant, the ownership of plant facilities, the number of customers to whom electricity is being sold, and the nature of the contract with those customers. The uncertainties surrounding the utility status of a cogeneration plant presents a significant barrier to cogeneration. Many, if not most, companies are reluctant to involve themselves in any cogeneration project where there is the slightest hint that they will be subject to PUC or FERC jurisdiction.

There are several ways of dealing with this uncertainty at the state level. First, in some states the PUC is authorized to issue advisory opinions on proposed projects. Industries may employ this route to resolve the question of state jurisdiction before they commit significant amounts of money to a cogeneration project. Alternatively, the PUC or its staff can issue a formal or informal pronouncement as to the extent of its jurisdiction over cogeneration facilities. The most direct way of dealing with the uncertainty is to amend regulatory statutes to clearly define the status of cogenerators. Such proposals are now pending before Congress[23] and the California State legislature, and another is being drafted by the Massachusetts Governor's Commission on Cogeneration.

What is the present regulatory status of industrial cogeneration projects? As has just been noted, this is difficult to say even in the context of a particular proposed project. In general, however, it appears that:

1.   An industry that generates electricity for use solely on its plant site and whose

transmission lines do not cross public property generally will not be subject to the jurisdiction of either FERC or the state PUCs. Where the plant's transmission lines cross public property, the industry may have to receive a local franchise and, in a few rare cases, may be subject to state PUC jurisdiction.

2. A cogeneration project that sells power to a select group of contract customers for their own use without holding itself out to serve the public generally will not be subject to PUC jurisdiction in the majority of states, nor to the jurisdiction of FERC. There is a substantial minority of states, however, in which the cogeneration project might be subject to PUC jurisdiction.

3. A cogeneration project that sells power to a public utility risks subjecting its plant to the jurisdiction of FERC. In a minority of states, it also risks subjecting itself to the jurisdiction of the state PUC (to the extent that FERC jurisdiction has not preempted state jurisdiction).

4. An industry that sells steam or heat to a utility for use in a cogeneration unit may be subject to PUC regulation in those few states that regulate the sale of heat or steam.

5. A cogeneration project that is owned jointly by an existing public utility and a private industry is likely to be subject to both state and federal regulatory jurisdiction.

A. Federal Energy Regulatory Commission Jurisdiction

FERC's jurisdiction is defined by the following statutory provision:

The provisions of this Part shall apply to the transmission of electric energy in interstate commerce and to the sale of electric energy at wholesale in interstate commerce, but shall not apply to any other sale of electric energy or deprive a State or State commission of its lawful authority now exercised over the exportation of hydroelectric energy which is transmitted across State lines. The Commission shall have jurisdiction over all facilities for such transmission or sale of electric energy, but shall not have jurisdiction, except as specifically provided in this Part and the Part next following, over facilities used for the generation of electric energy or over facilities used in local distribution or only for the transmission of electric energy in intrastate commerce, or over facilities for the transmission of electric energy consumed wholly by the transmitter.[24]

It is clear that an industrial company that owns or operates transmission facilities that are used to transmit electric energy to another company will be regulated under this provision. The limitation in this statute to "interstate commerce" has been virtually read out of the statute by the Supreme Court. Because of the interconnections between almost all utilities in the country, virtually every transmission of electric energy is an "interstate transmission". The more interesting and confusing question is whether an industry can avoid jurisdiction if it owns only generating equipment and not transmission equipment. The answer to this question hinges on the interpretation of the "but" clause in the above statute that provides that the FERC shall not have jurisdiction over "facilities used for the generation of electric energy". The only case directly interpreting that provision is *Hartford Electric Co. v. F.P.C.*[25] In that case, the FPC had issued orders to the Hartford Electric Company to comply with its uniform system of accounts. Hartford asked the 2nd Circuit Court of Appeals to overturn that order on the grounds that it was not a public utility because it owned only generation equipment and sold all of its power to another utility that owned all of the transmission facilities right up to the wall of the Hartford generating plant. The court rejected Hartford's plea and found it to be a public utility. Among other things, the court concluded that generation facilities were within the jurisdiction of the FPC when used as aids to the wholesale sale of electric energy. The court's precise arguments are too detailed and confusing to be discussed here. The reader is referred to the Dow study and the case itself for a detailed discussion.

It has been argued that the decision in a subsequent case before the Supreme Court (*Connecticut Light and Power v. F.P.C.*[26]) overruled the Hartford case by implication. Indeed, a very strong argument can be made that the Hartford case was wrongly decided. However, FERC has not shown any indication it feels the Hartford case is not sound law, and therefore, an industry that owns only generating equipment and sells power wholesale risks having FERC claim jurisdiction over its facilities. The only fashion in which the industry can clearly avoid FERC jurisdiction is to sell power directly to an end user. This, however, exposes the industry to possible regulation by state PUCs.

B. State Regulatory Jurisdiction

There is substantial variation among states as to who is subject to regulation by the public utility commission. In most states, the question of whether an individual person or corporation is to be subject to the jurisdiction of the PUC is synonymous with the question of whether that person or corporation is defined by state statute to be a "public utility". For example, in California the PUC may "supervise and regulate every *public utility* in the state and may do all things . . . which are necessary and convenient in the exercise of such power and jurisdiction."[27]

In the majority of state statutes, the term "public utility" refers to any entity which furnishes electricity in some manner to the public for compensation. Statutes have been generally interpreted by the courts to require that a company "dedicate its property to public use" before it will be held to be a public utility. Typically, a company will be found to have dedicated its property to public use where it is serving or has evidenced a readiness to serve an "indefinite public" which has the right to demand service. This does not mean that the company has to serve numerous customers to attain public utility status. In general, dedication to public use will not be found unless there is "substantial evidence of unequivocal intention" to dedicate private assets to the public. This intent may be found in acts which indicate a willingness to serve all who request service, in the wide solicitation of customers, in evidence that all customers who apply for service receive service, in voluntary submission to state regulation, or in use of the power of eminent domain.[6] In some states, the sale of electricity only to select contract customers has specifically been held not to constitute dedication to public use.[6] In other states, a company may sell power to a limited number of customers without being declared a public utility.

In states which follow this majority "dedication to public use" test, companies which generate power apparently will not be subject to regulation by public utility commissions where those companies consume that power themselves or sell it to a select group of contract customers.

However, a minority of states have taken the position that certain activities which do not involve dedication of property to public use are "so affected with the public interest" as to require regulation by the state public utility commission. This view has been used in at least two cases to find that shopping centers which generate their own electric power are subject to utility regulation. It is arguable that any industry cogenerating significant amounts of electricity so affects the public interest that it should be subject to the jurisdiction of the state PUC. This argument is least likely to succeed when directed toward an industry that cogenerates power only for its own use.

There are still other states whose statutes do not appear to tie regulation either to "dedication to public use" or to "affectation of the public interest", but rather, to the undertaking of certain specific actions such as the generation of electric power.

For example, the State of New York regulates "electric corporations" which it defines as corporations:

> . . . owning, operating, or managing any electric plant except where electricity is generated or distributed by the producer solely on or through private property for railroad or streetcar purposes or for its own use or the use of its tenants and not for sale to others.[28]

It would appear that under this statute a corporation cogenerating power would not be subject to state regulation if it consumed that power itself and without transmitting that power over public property. However, a corporation would seem to be subject to regulation if it sold or distributed *any part* of that power to another person or utility.

The status of companies selling electric power to public utilities is subject to even greater uncertainty than the status of companies consuming power themselves or selling power to select contract customers. Under the "dedication to public use" approach, a corporation may probably sell electricity to a public utility without state regulation if it does not hold itself out as willing to serve *all* public utilities in a given area. In states which follow the "affected with a public interest" approach, an industry is more likely to be declared a public utility when it sells power to a utility. In a number of states, there are statutory provisions which specifically grant public utility status to a company that sells electric power to a public utility. Note, however, that the states can regulate the wholesale sale of electric power only to the extent that it has not been regulated by the federal government.[6] Thus, if FERC has jurisdiction to set the wholesale electric rates charged by a cogenerator, the states will be preempted from controlling those rates, but will not be preempted from dealing with matters not touched by federal regulation, such as construction standards.

Uncertainty also surrounds three other likely cogeneration configurations: cogeneration plants jointly owned by an existing public utility and an industry, cogeneration plants where an industry sells steam or heat to a utility which in turn generates electric power, and centralized cogeneration plants where a private corporation supplies power to several industries within an industrial park.

As to joint ventures between an industry and a utility, the *Dow Study* concluded that:

> It seems possible that the state commissions having jurisdiction only over property dedicated to public use . . . would treat the public utility's interest in the venture corporation as an asset of the public utility and thus subject to regulation.[6]

Another commentator has concluded that all PUCs would probably regulate industry and utility joint ventures.[29] In general, it does appear that joint ventures between an industry and a *regulated* public utility will be subject to PUC jurisdiction in most states. An industry may, however, avoid jurisdiction by entering into a joint venture with a utility that is not subject to state PUC regulation. Municipal utilities, public utility districts, state utilities federal utilities, and cooperatives are often exempt from state regulation.

It has been suggested that utility ownership of the generating equipment in a cogeneration plant is an effective way of avoiding PUC jurisdiction over the private industry participating in a cogeneration arrangement. Under such an arrangement, the private industrial firm would sell steam or heat to a public utility which would use it to generate electricity. There is the risk under this arrangement, however, that the private industry will be subject to PUC regulation as a "public utility" engaged in supplying heat or steam. At least nine states now regulate steam-heating corporations.[30] These statutes could conceivably be applied to the private industry supplying steam or heat to a utility for power production.

Another likely configuration for cogeneration plants is a centrally owned and operated utility center in an industrial park. Under this configuration, a privately owned industrial center would cogenerate and sell steam and electricity to several industries located on a single industrial park site. At least one case has held that the industrial center is not a public utility under these circumstances.[31] The opposite result is highly likely in many states, especially where the industrial center submeters the electric power and steam rather than including it as part of the rent.[32]

When will an industrial cogeneration facility be subject to state or federal utility regulation? As this section has made clear, there is no easy answer to this question. Each PUC and state will have to look to its own statutes and case law, and even then it is likely that no clear answer will emerge for many likely cogeneration configurations.

How likely are PUCs to claim jurisdiction over cogeneration facilities? A study done by the National Research Council of the National Academy of Sciences examined the likelihood that one type of cogeneration system, the Modular Integrated Utility System (MIUS), would be subject to state regulation. The following conclusion was reached:

In each of the four areas, studied, the staff of the regulatory agency expressed the opinion that the agency would seek to regulate a MIUS and would be opposed to an unregulated facility. Nonetheless, the agency's authority to regulate a MIUS is uncertain.

Even if MIUS is not subject to regulation as a public utility, the opposition of the regulatory agency may make it difficult to obtain the needed permits and approvals from the other state and local agencies. Such agencies frequently seek and follow the advice of the regulatory agency.[33]

C. Legal Requirements Affecting the Operation of a Cogeneration Plant

There are numerous other laws that will need to be considered during the design and implementation of a cogeneration facility. Many of these laws are unique to a particular state or region. Thus, each jurisdiction should examine the full spectrum of its statutes to see which laws will impact the implementation and operation of cogeneration and which of those need to be altered to accommodate cogeneration. A few of the more common types of legal obligations affecting cogeneration plants are outlined briefly below.

1. Power Plant Siting Statutes

At least 23 states have enacted power plant siting statutes that require a state commission, council, board, or department to approve the siting of large energy facilities.[34]

All states that have or are actively considering facility siting statutes should examine that possible impacts of their laws on the implementation of industrial cogeneration. Where possible, exemptions for small cogenerators should be considered.

2. Fuel Allocation Priorities

Under existing fuel allocation priorities that will be used in the event of an emergency fuel shortage, industrial users are given a lower priority than utility users.[35]

3. Air Pollution Controls

As noted above, cogeneration can reduce the discharge of air pollutants overall, but it does so at the cost of increasing emissions from particular industrial plants. An industry that will emit increased air pollutants by reason of operating a cogeneration plant may have difficulties meeting applicable pollution control standards, especially if it is located in a nonattainment area, an area that is not in compliance with national ambient air quality standards.

4. Personnel Requirements

Most local codes set minimum requirements for those persons who are operating high pressure steam systems.

VII. POLICY OPTIONS FOR LIFTING THE BARRIERS TO COGENERATION

A. State Initiatives

There has been relatively little activity at the state level toward overcoming the barriers to industrial cogeneration. A few states have held preliminary discussions or workshops on the subject. The governors of New Jersey and Massachusetts have appointed commissions to study possible policy initiatives in their states. The California State Energy Commission held a one-day workshop bringing together industrial, utility, and government leaders. Academic studies of cogeneration have been commissioned by Maine, New Jersey, California, and New York. Of all these efforts, the most significant steps to encourage cogeneration have been taken by the California PUC and the state of Massachusetts.

The California Public Utilities Commission has for several years been in the forefront of those seeking to encourage waste-heat utilization and industrial cogeneration. Its staff has been working actively with the state's utilities to develop regulations and utility rates that support implementation of cogeneration projects. On March 16, 1976, the Commission issued an order that required utilities to consider options for waste-heat utilization, including industrial cogeneration. When this order failed to produce sufficient progress, the Commission on January 10, 1978 issued a much stronger order requiring that the state's three major electric utilities:

1. Provide specific rate proposals for enhancing cogeneration, including revised standby rates, and new interruptible electric service rates that would permit industry to obtain standby service on an interruptible basis at lower rates than regular standby rates
2. Fully identify long-term cogeneration potential within their service area and provide schedules for bringing these potential cogenerators on-line
3. File status reports on cogeneration projects being considered and update those reports quarterly
4. Submit guidelines covering the price and conditions under which the utilities would purchase energy and generating capacity owned by industry or other private power generators
5. Submit guidelines for the development of utility-owned cogeneration facilities.[23]

In addition to helping utilities implement this order, the PUC staff has assigned a senior-level engineer full time to develop costs for standby service and energy purchased by utilities, to develop cost-based rates for standby service and price guidelines for the purchase of cogenerated power, to assist utility customers in the review of cogeneration projects, and to develop project criteria that will expedite cogeneration projects.

The California utilities have begun complying with this order. Southern California Edison Company has developed an interruptible-time-of-use industrial rate schedule that enables industry to purchase interruptible standby power for less than normal standby power rates. In addition, it has developed three alternative contract formats under which it will buy power from private sources or contract to operate a cogeneration facility in cooperation with a private company. Pacific Gas and Electric Company has responded to the order by developing a new standby rate schedule that allows a customer to have its standby charges reduced based on the demonstrated reliability of its self-generation system. PG&E has also prepared guidelines stating how it will calculate the price it will offer for privately generated power. All three California utilities are now negotiating with numerous industries for the establishment of cogeneration capacity, and are examining the possibility of constructing their own cogeneration plants.

Other important steps toward removing the barriers to cogeneration have been taking place in Massachusetts. The governor of that state has appointed a commission to study the subject and recommend policy initiatives. As of June 1978, the Commission had not yet issued its findings or recommendations. The staff, however, was considering the following proposals:

- An industry should not be subject to the jurisdiction of the Department of Public Utilities (DPU) unless it has a capacity of 300 MW or greater and serves 50 customers or more.
- A nonutility-owned cogenerator that is subject to DPU regulation should be exempt from full-scale rate hearings and fuel adjustment proceedings where all of that company's rates are set by private contract.
- An industry that is subject to DPU regulation should be required to file an annual report on only that portion of its business related to electricity sales.
- Energy conservation should be included under the DPU's concept of the "public interest", and therefore, limited competition should be permitted in the generation of electricity.
- Cogenerators should be permitted by the PUC to petition for backup power.
- Cogenerators should not be required to sell power to persons demanding such a sale.
- When hearing rate cases, the DPU should carefully examine the extent to which the utility requesting a rate increase took advantage of cogenerated electricity and the price it paid for any cogenerated power it purchased.
- In passing on contracts for the purchase of cogenerated power, the DPU should balance the interest of the utility customers in reasonable rates, the interest of the cogenerator in a fair price, and the interest of the public in the conservation of resources and efficient use of energy.
- In reviewing time-of-day electric rates and utility-load management programs, the DPU should insure that cogenerators are offered fair and reasonable rates for their power and that the utilities have adequately considered it as a factor in reducing their load.
- DPU safety standards should be applied to cogenerators.
- The DPU should have authority to order interconnection and wheeling for the limited purpose of permitting an industry to sell power at wholesale and to enable the utility to satisfy its obligations under the regional power-pool agreement.
- The regional power-pool agreement should be amended to reduce the minimum size of the power producers with which a utility is permited to contract.
- The state's facility siting statute should be amended to apply only to cogeneration units having a capacity of 300 MW or greater and having 50 customers or more.
- The state energy facility siting council should require utilities to consider cogeneration as a generation alternative in their long-range forecasts.[36]

B. Federal Initiatives

There has been a rapidly growing interest in industrial cogeneration at the national level. The federal government continues to fund a series of studies on the subject and has established an interagency task force to coordinate federal programs. FERC has been examining its role in cogeneration. Major utility groups, including the Electric Power Research Institute and the Edison Electric Institute, have undertaken examinations of the concept. Conferences on cogeneration are cropping up across the country, and there is now even a cogeneration society.

President Carter's National Energy Plan came out strongly in support of increased cogeneration. It proposed insuring that industrial firms receive a fair price for the power they sell to utilities and for the backup power they must purchase. In addition, the National Energy Plan proposed permitting the Department of Energy to exempt industrial cogenerators from federal and state regulations, provided an additional 10% investment tax credit for industrial investments in cogeneration equipment, and provided that industrial cogeneration facilities could be exempted from coal-conversion requirements.

It now appears highly likely that there will be federal legislation on cogeneration in the near future, although the precise outlines of that legislation are still cloudy. The congressional conference committee considering the National Energy Act has reportedly reached the following agreement. FERC will be required to prescribe rules requiring electric utilities to offer to sell or buy power from qualifying cogenerators. Those rules shall insure that the rates set by State Regulatory Commissions for these sales will be subject to specified limits and will not discriminate against cogenerators. FERC will also be empowered to prescribe rules exempting qualifying cogenerators and small power producers of up to 30 MW from state utility regulations, the Federal Power Act, and the Public Utility Holding Company Act if the Secretary of Energy determines that such exemption is necessary to carry out the purposes of the act. A qualifying cogenerator will be defined as the owner of a cogenerating facility that meets size, fuel use, and fuel efficiency standards prescribed by FERC.[40]

The prospect of federal legislation should not, however, serve as an excuse for inaction at the state level. The barriers to cogeneration are too numerous, too complex, and too localized to be efficiently dealt with entirely at the federal level. If cogeneration is to expand significantly, active state participation in removing the institutional barriers is critical. In fact, all the federal legislative proposals to date have recognized this fact by providing that the federal government can delegate many of its powers under the proposed statutes to a state agency.

NOTE ADDED IN PROOF

Since the preparation of this manuscript, Congress has enacted the Public Utility Regulatory Policies Act of 1978, Pub. Law 95-617 (November 9, 1978). That act grants FERC significant powers to mitigate the legal and regulatory barriers to industrial cogeneration. Detailed descriptions of the provisions of the new federal law are contained in Reference 37.

REFERENCES

1. **Anon.**, Draft Report of the Governor's Commission on Cogeneration, Boston, Mass., June 8, 1978.
2. **Murray, F. X.**, *Where We Agree: The Report of the National Coal Policy Project*, Vol. 1, Westview Press, Boulder, Colo., 1978.

3. **Williams, Robert H.,** *Industrial Cogeneration,* Center for Environmental Studies, Princeton, N. J., 1978.

4. Resource Planning Associates, *The Potential for Cogeneration Development in Six Major Industries by 1985,* Resource Planning Associates, Cambridge, Mass., 1977.

5. **Williams, Robert H.,** *The Potential for Electricity Generation as a By-Product of Industrial Steam Production in New Jersey,* Center for Environmental Studies, Princton, N. J., 1976.

6. Dow Chemical Company, *Energy Industrial Center Study,* Dow Chemical Co., Midland, Michigan, 1975.

7. 1975 Pennsylvania Laws, Act No. 57, §5(a).

8. *Kentucky Utilities Co. v. Public Service Commission,* 252 S. W. 2d 885 (Ky. Ct. App. 1952).

9. **Jones, William K.,** *Regulated Industries: Cases and Materials,* Foundation Press, Brooklyn, N. Y., 1976.

10. **Ferris, M. T. and Sampson, R. J.,** *Public Utilities: Regulation, Management, and Ownership,* Houghton Mifflin, Boston, 1973.

11. *Oklahoma Gas and Electric Co. v. Total Energy, Inc.,* 499 P. 2d 917 (Okla. Sup. Ct. 1972).

12. **Priest, A. J. G.,** *Principles of Public Utility Regulation,* Vol. 1 and 2, Michie Co., Charlottesville, Va., 1969.

13. *Kornegay v. City of Raleigh,* 269 N.C. 155, 152 S.E. 2d 186 (1967).

14. Oklahoma Constitution Art. §5(a).

15. Wisconsin Stat. Ann. §196.58(4) (1957).

16. Staff Report on California Cogeneration Activities, California Public Utilities Commission, San Francisco, Calif., January 17, 1978, 13.

17. **Johnson, Kenneth E.,** *Possibilities for Electricity-Steam Cogeneration Systems in Maine,* Energy and Environmental Policy Center of Harvard University, Cambridge, Mass., 1977.

18. Southern California Edison Company, Schedule No. S, Standby, at California Public Utilities Commission Sheet No. 3598-E, California Public Utilities Commission, San Francisco, July 1, 1964.

19. **Murray, Francis X.,** Ed., *Where We Agree: the Report of the National Coal Policy Project,* Vol 1, Westview Press, Boulder, Col., 1978, 228.

20. Amended application No. 55509 before the Public Utilities Commission of the State of California, San Francisco, undated.

21. California Public Utilities Commission, Order Directing Electric Utilities to Augment Cogeneration Project, Resolution E-1738, January 10, 1978, San Francisco, Calif.

22. Pacific Gas and Electric Company, Response to Item No. 5 of the CPUC Order No. E-1738, on file at California Public Utilities Commission, San Francisco, March 13, 1978.

23. H.R. 4018, 95th Cong., 1st Sess., 1977; H. R. 8444, 95th Cong., 1st Sess., 1977.

24. 16 U.S.C. § 824(1974).

25. *Hartford Electric Light Co. v. F.P.C.,* 131 F.2d 953 (2nd Cir. 1942), *cert-denied,* 319 U.S. 741 (1943).

26. *Connecticut Light and Power Company v. F.P.C.,* 324 U.S. 515 (1945).

27. California Public Utilities Code § 701 (1975). (Emphasis added.)

28. N. Y. Public Service Law 2(13) (1955).

29. **Uhler, Lewis K.,** Law and regulation of the electric utility industry: toward expanded public choice, in *Electric Power Reform: The Alternatives for Michigan,* Shaker, William H., Ed., University of Michigan Institute of Science and Technology, Ann Arbor, Mich., 1976, 301.

30. National Association of Regulatory Utility Commissioners, 1974 Annual Rep. on Utility and Carrier Regulation, Washington, D.C., 1976, 368.

31. *General Split Corp. v. P. & V. Atlas Industrial Center, Inc.,* 44 P.U.R. 3d 334 (1962).

32. *See Re Boston Edison Co.,* 98 P.U.R. (NS) 427 (1953).

33. **Anon.,** Evaluation of Integrated Utility Systems, National Academy of Sciences, U.S. Department of Housing and Urban Development, Washington, D.C., 1975, 59.

34. **Anon.,** *Power Plant Siting in the United States,* Southern Interstate Nuclear Board, Atlanta, Ga., 1976.

35. Draft material supplied to the author by the staff of the Massachusetts Cogeneration Commission.

36. Summary of the Conference Committee Agreement on the Public Utility Regulatory Policies Act, U.S. House of Representatives, December 1, 1977, unpublished.

37. Environmental Law Institute, *Energy Efficiency in Industry,* Ballinger Publishing, Cambridge, Mass., in press.

Chapter 14

COGENERATION AND DISTRICT HEATING SITING

B. W. Wilkinson

TABLE OF CONTENTS

I. INTRODUCTION

A significant concern in the decision to install a cogeneration/district heating project is the location of the power plant. For industrial situations where the users of the by-product heat are already in place, the availability of power plant sites may be dictated by the available space. As will be discussed below, a maximum steam transmission distance exists beyond which the energy and economic advantages of a cogeneration system are offset by the penalties of capital cost and heat losses associated with the pipelines. Thus, for industrial plants, it is often easier to reach a conclusion about a cogeneration site simply based on the distance to available plant sites. District heating, on the other hand, is somewhat more complex with regard to siting. Once again, the plant site needs to be reasonably close to the service area, but perhaps an even more important criterion is the layout and location of the heat distribution system. This must be done with particular care in existing cities. This in itself may dictate the location of the heat source.

In the material presented below, an example is given of the economic comparison of the heat (steam) transmission distances vs. the economies of scale which are associated with larger, centralized power sources.

A further example is given of the methodology that can be used to evaluate the possible locations for industrial cogeneration sites. In fact, the potential sites for large nuclear plants which might supply process steam are identified by this technique.

II. ECONOMIC PLANT LOCATIONS

The optimum location of a power plant may be determined by balancing the capital costs and heat losses associated with the transmission pipeline with the capital advantage associated with the construction of a larger power plant capable of supplying multiple industries as compared to the construction of small power plants at each industrial site. In one such comparison,[1] various power plant locations were analyzed for industrial complexes, each of which required from 0.5 to 1 million lb/hr of steam. Several industry geometries were chosen. In one case, a two-industry comparison of the cost of energy at each industrial site was made with a cogeneration plant supplying both steam and electricity to both industries and located midway between the industries (see Figure 1). The independent variable for this study was the distance between the two industrial plants. The algorithm calculating the total cost of energy to the industries produced the results shown in Figure 2. As may be seen, the centralized cogeneration case is more economical at distances between the two industries of up to about 7.5 mi. Increasing the steam load to the industries to 2 and 4 million lb/hr, however, does allow an increase in the spacing of the plants since larger steam lines can be constructed and amortized over a larger amount of steam, and the larger power plant shows an additional economy of scale. Thus, the maximum distance of 7.5 mi calculated above may be assumed to expand to 10 to 20 mi with increasing loads. This is consistent with a similar study[2] which assumes that industry steam needs of 1.0 million lb/hr can be economically met with a single power plant as long as the plants are located within a 2 mi radius.

Consistent with this reasoning, a two-industry study[1] was expanded to three industries. For simplicity, the industries were assumed to lie equidistant from each other as shown in Figure 3. Again, the results of the study[1] show that the centralized power plant is more economical than the decentralized (individual plant) system at distances up to 9 mi (see Figure 4).

When the plant grouping is expanded from three to six industries, the maximum distance between plants further expands to 13 mi.

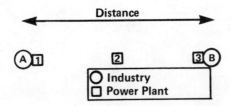

FIGURE 1.   Generalized two-industry case.
(From Tiller, J. S., Wilkins, R. D., and Hilsen, N. B., *Proc. Southeastern Region 3 Conference,* Institute of Electrical and Electronic Engineers, New York, 1977, 568. With permission.)

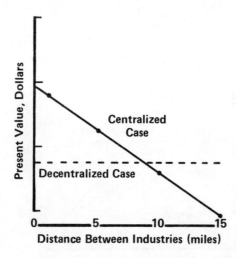

FIGURE 2.   Results of two-industry study. (From Tiller, J. S., Wilkins, R. D., and Hilsen, N. B., *Proc. Southeastern Region 3 Conference,* Institute of Electrical and Electronic Engineers, New York, 1977, 568. With permission.)

The determination of suitable plant-site locations has been investigated for nuclear plants.[3] The strategy for nonnuclear heat sources would be identical since, even at the largest industrial site locations, the steam consumption is a small fraction of the total energy output of a power plant. Thus, smaller coal-fired plants would still be capable of supplying sufficient energy.

In order to evaluate the existence of plant sites of suitable size, it was decided to limit the study to those locations which met one or more of the following criteria:

- Locations which have a combined industrial steam requirement of at least 500,000 lb/hr within a circle of 4 mi diameter
- Locations which have a combined industrial steam requirement of at least 2,000,000 lb/hr within a circle of 10 mi diameter
- Locations which have a combined industrial steam requirement of at least 4,000,000 lb/hr within a circle of 20 mi diameter

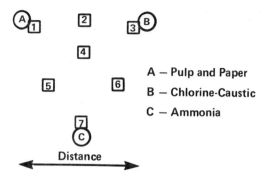

FIGURE 3.   Generalized three-industry case. (From Tiller, J. S., Wilkins, R. D., and Hilsen, N. B., *Proc. Southeastern Region 3 Conference,* Institute of Electrical and Electronic Engineers, New York, 1977, 568. With permission.)

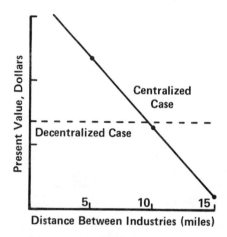

FIGURE 4.   Results of three-industry study. (From Tiller, J. S., Wilkins, R. D., and Hilsen, N. B., *Proc. Southeastern Region 3 Conference,* Institute of Electrical and Electronic Engineers, New York, 1977, 568. With permission.)

Sites which had either smaller loads or which had more widely dispersed users were not considered to be economically acceptable.

In order to identify the locations of significant process heat, an analysis of process demand is necessary. Industrial process heat requirements represent approximately 25% of the total demand for combustible fuels (excluding feedstocks).[4] One half of this is used for the production of steam. The balance is used in open-fired kilns, driers, furnaces, etc. These require, in general, higher temperatures than for steam generation. Most of the process steam applications are at temperatures from 100 to about 450°F. Approximately 85% of the industrial steam requirement is below 400°F and within the range available from conventional nuclear reactors.

Several industrial classifications were found to be the principal consumers of energy for process use. A previous study showed that the following SIC groups consumed the fraction of industrial energy shown (data for 1973) in Table 1.

TABLE 1

Industrial Energy Con-
sumers

Consumer	%
Primary metals	20.4
Chemicals	21.4
Petroleum	12.5
Pulp and paper	4.9
Stone/glass/clay	4.6
Others	36.2
Total	100.0

Since the first four categories represent about 60% of the energy consumed by industry, these were selected for further investigation. Using literature sources, the locations, products, and capacities of industrial plants were determined. Unit ratios for the various products were estimated by the published steam consumption for representative processes. Published energy consumptions varied rather widely, depending upon the process flow sheet chosen. Wherever possible, knowledge of the process being used established the proper choice of unit ratio for steam. If such information was not available, a conservative ratio was assumed in the calculation of the total steam demand for a particular plant site. Representative steam unit ratios are shown in Table 2.

The results of these calculations were checked (where possible) with data supplied for the plant in question in a previous study.[4] Good agreement was found.

The exact location of each plant was identified by the personal knowledge of personnel familiar with the area, or from published information available for several high-concentration areas and by direct telephone calls to many of the industrial plants concerned.

A final check on both steam loads and plant locations was achieved by mailing a copy of the results for each plant location directly to the plant for verification. Responses were received from over 70% of the addresses, and the data were corrected where it was necessary to do so.

The results of the study (see Figure 5) indicate that there are at least 110 locations in the U.S. which have a combined industrial steam load of at least 1,000,000 lb/hr within a circle of 4 mi in diameter. An additional 24 locations have a combined steam load of at least 2,000,000 lb/hr within a circle of 10 mi diameter. Locations which have a combined steam load in excess of 4,000,000 lb/hr within a circle of 20 miles diameter amounted to 19 additional sites.

It is believed that the steam requirements are conservative for each location. Only the four specified industries (plus the rubber industry) were examined, and individual plants using less than 200,000 lb/hr were not considered. Thus, individual sites may have additional steam requirements.

In the analysis of potential sites, total steam demand was the principal factor. It should be noted that some of this steam is produced by process residuals or by-product material suitable as fuel. Examples of such materials are the black liquor and wood waste in the paper industry, certain refinery gases of the petroleum industry, and blast-furnace gases of the iron and steel industry. Estimates of the fraction of total steam demand that could be produced from this source were made for each industry involved. In general, these amounted to 25 to 50% of the total steam demand. Since the amount

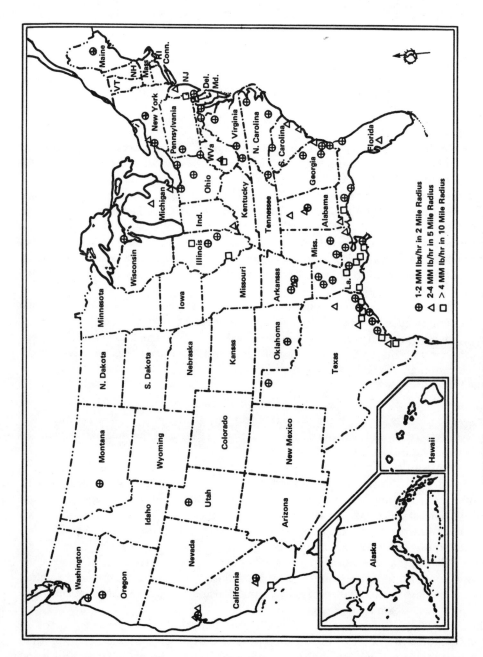

FIGURE 5.   Potential industrial steam/power cogeneration sites.

TABLE 2

Typical Steam Unit Ratios

Product	Steam consumption	lb/Unit product
Low density polyethylene	0.8	lb/lb
High density polyethylene	2.8	
Polystyrene	0.4	
Polyvinylchloride	0.5	
Methyl alcohol	3.0	
Phosphoric acid, wet process	0.7	
Petroleum refining	5.0	lb/hr/BPD

of process residuals which are produced and used within any given plant is sensitive, proprietary information, it is not possible to define this factor precisely, and thus, the estimates of replaceable steam (not produced by residuals) is likely to be within the range of 50 to 75% of the total steam load calculated.

The next reasonable step in site evaluation would be the evaluation of an acceptable location within the industrial region in question. Among the criteria examined for suitability, the requirements of 10 CFR Part 100[5] and NRC Regulatory Guide 4.7[6] would be applicable for nuclear plants. Obviously, questions of zoning as well as geologic suitability would be applicable to any plant selected. As mentioned earlier, the most reasonable site for the power plant would probably be equidistant from the industrial plants in question.

REFERENCES

1. Tiller, J. S., Wilkins, R. D., and Hilsen, N. B., Determination of economically attractive locations for steam and electric cogenerating plants, Proc. Southeastern Reg. 3 Conf., Institute of Electrical and Electronic Engineers, New York, 1977.
2. Energy Industrial Center Study, Grant No. OEP74-20247, prepared for the Office of Energy Research and Development Policy, National Science Foundation, Washington, D.C., June 1975.
3. Wilkinson, B. W. and Barnes, R. W., Dual Purpose Nuclear Plants: Potential Industrial Sites, Proc. 5th Conf. Energy and Environment, American Institute of Chemical Engineers, Dayton Section, Dayton, Ohio, 1978.
4. Barnes, R. W., New Energy Sources for Process Heat, NSF-OEP74-18055, National Science Foundation, Washington, D.C., 1975.
5. Code of Federal Regulation, Title 10, Part 100, Reactor Site Criteria.
6. Nuclear Regulatory Commission Guide 4.7, General Site Suitability Criteria for Nuclear Power Stations, U.S. Nuclear Regulatory Commission, Washington, D.C., 1975.

Chapter 15

POTENTIAL FUTURE MARKET FOR INDUSTRIAL COGENERATION

Peter Bos and James Williams

TABLE OF CONTENTS

I. INTRODUCTION

With rising energy costs and dwindling energy supplies, cogeneration has become an increasingly attractive alternative to conventional power generation for industry. However, a number of technical, economic, environmental, and institutional constraints will prevent full development of industrial cogeneration. To assess how these constraints would be minimized, Resource Planning Associates, Inc. (RPA) and a subcontractor, Burns and Roe, Inc., carried out a study for the U.S. Department of Energy to assess constraints to cogeneration and the effects of alternative government actions to stimulate cogeneration.[1] Since initial analysis indicated that 60 to 70% of potential development is in the area of process steam topping, the study focused on that area. Further, the study considered six industries that account for a major percentage of all electricity and process steam used in the industrial sector: chemicals, petroleum refining, pulp and paper, steel, food processing and textiles. Only applications involving electrical output of more than 5 MW were studied, because these offer the highest potential industrial energy savings. In addition, the study considered only industrially proven technologies and set 1985 as the appropriate time horizon.

According to the RPA study, the total amount of cogeneration development by 1985 would depend on the action taken by the federal government and other interested parties. From the industrial investors' viewpoint, without special government support, only a total of about 1250 trillion Btu of process steam for those six industries will be produced by additional cogeneration plants installed over the 1977 to 1985 period. This process steam will be produced by approximately 6000 MW of additional (cogeneration) electric power capacity. With combined federal actions* directed at removing economic and regulatory constraints, an additional 2000 to 6000 MW of electricity could be cogenerated in those industries, which would result in gross annual energy savings of up to 300 trillion Btu**, or almost 150,000b/day oil equivalent. Although not estimated, potential savings could be significantly higher in the long term, (i.e., 1985 to 2000) because of (1) a greater percentage of new (and more economically attractive) industrial facilities and (2) more time available for government, utility, and industrial actions to take effect.

Somewhat similar studies of cogeneration potential have been completed by other groups.[2-9] The range of results is indicated in Table 1,[10] and the variation in results shown is due, in part, to the extent to which the various technical, economic and legal constraints have been factored into the studies.

The greatest potential for cogeneration development lies in a coordinated approach by decision makers in each of the four major interacting sectors: the federal government, state governments, industry, and electric utilities.

In the following document, the potential for industrial cogeneration development is discussed:

- Without federal action
- With federal action
- With coordinated action by the four major interacting sectors.

---

* Individual government actions are discussed in the next section.
** Gross energy savings do not reflect increased waste fuel consumption.

TABLE 1

Estimates of Future Energy Conservation from Cogeneration Quadrillion Btu/yr (Quads)[a]

	Indefinite future, based on present energy use			1985			2000/2010		
	A	B	C	A	B	C	A	B	C
Industrial cogeneration (Ref.)									
2					1.5				
3		1.2—4.6		1.7—6.9	0.6—5.2				
4						1.5[b]			
5							6—10		2—4
6			2.6						
1[c]				1.1—1.9	0.5—0.9[b]	0.5—0.7[d]			
Utility central station cogeneration (district heating) (Ref.)									
7		3.3—3.8							
8		4.7			5.2			5.0	
Modular integrated utility systems (MIUS) (Ref.)									
6			0.9						
9									0.3—0.6[e]

Note: A, Technological potential; B, technological and ecological potential; C, planned or expected actual.

[a] Fuel reduction compared to meeting the same heat and electricity demands without cogeneration. Consideration is given to steam generation only (excludes bottoming cycles for waste-heat recovery and also topping cycles on direct process heat).
[b] With government action.
[c] RPA figures have been increased to include 0.16 Quads of energy saving from cogeneration facilities existing in 1975.
[d] Without government action.
[e] Residential sector only.

II. COGENERATION DEVELOPMENT WITHOUT SPECIAL GOVERNMENT ACTION

More than 1700 trillion Btu of process steam (representing about 450 trillion from existing cogeneration plants and about 1250 trillion from new cogeneration plants), and almost 80 billion kWh of electric energy, are likely to be cogenerated by 1985 in the chemical, petroleum refining, pulp and paper, steel, food processing, and textiles industries. This represents about 10,000 MW of cogeneration capacity, compared with a current operating level of somewhat more than 4000 MW.

There are three important markets for cogenerated steam. There is the *near-term retrofit market* which represents the cogeneration potential at manufacturing facilities with existing process steam generation capacity where cogeneration is currently technically and economically feasible; the *mid-term retrofit market* which represents the cogeneration potential at manufacturing facilities with existing process steam generation capacity, but without the ability to install cogeneration capacity for technical, economic, environmental, regulatory, or institutional reasons; and the *new development market* which represents the cogeneration potential at new facilities that will require steam generation capacity from 1977 to 1985.

To project demand for cogenerated steam in these markets, a number of factors must be considered. First, of the total steam demand, steam technically suitable for cogeneration must be determined. Of that quantity, process steam demand not eliminated by declining steam demand and high steam-to-electric demand considerations must be distinguished. Finally, financially attractive cogenerated steam must be isolated from the remaining total. By eliminating steam that does not meet these criteria, a total cogeneration potential can be estimated.

A. Technically Suitable Steam

Of the 5900 trillion Btu of process steam projected to be used in 1985, about 3500 trillion Btu is technically suitable for additional cogeneration development. Conversely, more than 40%, or about 2400 trillion Btu, is either technically unsuitable for supply by a cogeneration plant or will be supplied from existing and planned cogeneration plants and mechanical-drive noncondensing systems (see Figure 1).

The major limiting technical factors are low pressure steam from waste-heat-recovery boilers, out-of-phase steam-electric load fluctuations, low steam demand, and existing contracts for purchased steam. If a plant has an opportunity to employ a low pressure waste-heat-recovery boiler, it will use that waste heat instead of cogenerated steam.

Low pressure waste-heat-recovery boilers are estimated to produce about 8% (or approximately 540 trillion Btu of process steam) in 1985, which represents a direct reduction in the potential cogenerated steam. On one hand, waste-heat-recovery boilers are now more financially attractive at existing and new manufacturing facilities. On the other hand, energy-saving technologies will be implemented at new manufacturing facilities to save waste-heat losses, which could reduce both the input process steam needed and the "bottoming" cogeneration potential.

Out-of-phase steam-electric load fluctuations will reduce the cogeneration steam potential by about 7% (or approximately 425 trillion Btu). At new manufacturing facilities, however, this technical constraint might be reduced if technological innovations related to load smoothing were developed.

If steam demand is too low in a manufacturing plant, that plant cannot be practically served by existing cogeneration technology. For the six industries examined, this constraint will reduce the cogeneration potential by about 5% (or approximately 365 trillion Btu of process steam) in 1985.

Finally, process steam is purchased by many manufacturing plants. While these contracts remain in effect, this steam cannot be considered for cogeneration capital investments. For these six industries, this constraint represents an estimated reduction in cogeneration potential in 1985 of 2% (or about 140 trillion Btu of process steam).

B. Steam Considered for Capital Investment

By 1985, the amount of process steam which will be considered for cogeneration investment is about 41% of the total process-steam demand (or approximately 2400 trillion Btu). Two additional factors will further reduce process-steam cogeneration potential by 19% (or about 1100 trillion Btu) in 1985, (1) expectations of a declining steam demand and (2) high steam-to-electric demand conditions.

Anticipated process changes or plant shutdowns and steam conservation programs will decrease potential cogeneration development by about 8% (or approximately 460 trillion Btu) in 1985 (see Figure 2). Since the planned life of a cogeneration plant is 20 years or more, such plants would generally not be built for any manufacturing steam demand that would be scheduled for phaseout within a shorter time frame. Furthermore, the reluctance of industrial companies to invest in cogeneration in high steam-to-electric demand situations, which require sizing the cogeneration plant to the manufacturing facility steam load and selling the excess power to the utility system (possible

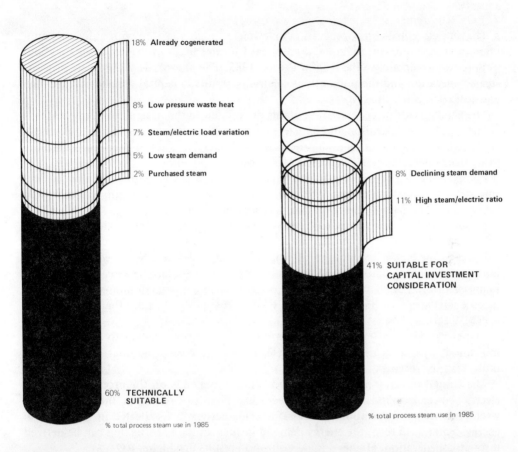

FIGURE 1. Technical constraints to cogeneration in six major industries.

FIGURE 2. Preinvestment knockouts.

under government regulations) will further reduce the cogeneration potential by about 11% (or approximately 655 trillion Btu) by 1985.

Thus, of the total 1985 process steam demand of 5900 trillion Btu, approximately 59% (or 3500 trillion Btu) is estimated to be unsuitable for cogeneration capital investment consideration.

C. Financially Attractive Cogeneration

Without government action, about 1250 trillion Btu of process steam is expected. Of this amount, 250 trillion Btu is currently technically and economically suitable for development, and 1000 trillion Btu is projected to be both technically suitable and financially attractive enough for industry investment in new cogeneration plants by 1985. The most financially attractive cogeneration configurations to decision-makers in the six industries would provide approximately 6000 MW of additional cogeneration capacity.

Of the 2400 trillion Btu of process steam suitable for capital investment consideration in 1985, almost 1400 trillion Btu of potential cogeneration is financially unattractive because cogeneration investments often do not meet industry's return on investment (ROI) requirements, and because of unacceptable risks associated with cogeneration plant investments, such as uncertainty over fuel price and availability, environmental standards, and government regulations (see Figure 3).

1. ROI Requirements

The limited returns on cogeneration investment, in comparison with industry's minimum acceptable ROIs, reduce the potential for cogeneration development by about 22% (or approximately 1300 trillion Btu) in 1985. The alternative of separate process-steam generation and the purchase of electricity results in capital costs which are high and cost savings which are low.

The capital costs of cogeneration plants are related to the design requirements, that is, the type of configuration and its size. Obviously, industrial cogeneration plant boilers have higher capital costs than process steam boilers because of the need for a more expensive boiler, emission controls, and electrical equipment.

For example, a 30 MW coal-fired system with a straight back-pressure nonconden-sing steam turbine could cost $40 million, about two thirds more than a coal-fired process steam boiler with the same steam-generating capacity.

In addition, the investment requirements are disproportionately high for small cogeneration plants, thereby producing low ROIs for these plants. Although this is attributed to the lack of economies of scale for electric devices, boiler equipment, and most system components, one of the major investment requirements is the steam turbine-generator. For instance, a 5 MW noncondensing steam turbine-generator with turbine inlet steam of 850 psig and 825°F has a cost per MW more than twice that of a 30 MW turbine.

The cost savings offered by cogeneration plants vary according to plant capacity utilization, fuel costs as compared to electric energy costs, and charges for electric utility standby services.

The annual utilization of cogeneration capacity depends on the process steam and electric power demand of the manufacturing facility throughout the year. Daily, weekly, or seasonal variations in manufacturing activity are reflected in variations in energy demand. In turn, the energy demand influences the benefits that can be derived through cogeneration. Higher annual utilization results in a higher ROI.

High fuel costs and low electric power rates also reduce the cost savings of a cogeneration plant. Because cogeneration enables industry to replace purchased electrical power with cogenerated power, a cogeneration plant is more attractive when the electric rate is high and the cost of industrial fuel is low. Because a cogeneration plant will consume more fuel for its dual power outputs than a boiler generating only the equivalent amount of process steam, the incremental fuel cost required to produce the cogenerated electric energy can be sufficiently high and the purchased electric cost can be sufficiently low to make the ROI unattractive.

High charges for utility standby service when the industrial facility is unable to generate the required electric power can also reduce cost savings for a cogeneration plant.

2. Uncertainties and Risks

Management's perception of the risk and uncertainties of economic factors, environmental controls, and regulatory and institutional constraints surrounding a potential cogeneration project constrains cogeneration investment by raising the required ROIs. Four major factors limit cogeneration development, (1) capital availability, (2) fuel availability and price, (3) environmental policy and control equipment, and (4) government regulations. In combination, these risks would reduce cogeneration potential by almost 2% (or about 100 trillion Btu) in 1985.

Financial/Economic Constraints

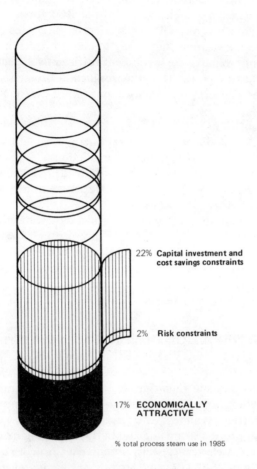

22% Capital investment and
cost savings constraints

2% Risk constraints

17% ECONOMICALLY
ATTRACTIVE

% total process steam use in 1985

FIGURE 3. Financial/economic constraints.

a. Capital Availability

The capital investment required for a cogeneration plant is considerable. Many companies do not have sufficient internally generated funds to finance the expenditure, and their access to capital markets may be limited. The necessary investment for a cogeneration plant could increase the company debt to an unacceptable level. In addition for some companies, the preference for market-oriented investments over cost-saving projects further limits the acceptability of cogeneration-related investments.

b. Fuel Availability and Price

Managers are concerned that the fuel used in a cogeneration facility may become scarce or very costly. Management is concerned about future curtailments in the supply of natural gas and oil, the potential effects of the deregulation of oil and gas prices, the lack of price stabilization of coal, and the future capability of the coal mining and transportation industries to meet demand.

In view of these uncertainties, some managers are reluctant to increase fuel demand by installing new or additional cogeneration facilities. In fact, despite rising electricity prices, some managers might switch from steam-powered to electrical-powered equipment, thus reducing the demand for steam and, hence, the potential for cogeneration.

c. Environmental Policy and Control Equipment

The potential cost and technical reliability of environmental controls have caused some managers to demand a higher-than-normal ROI for investments involving increased environmental emissions. Other managers have rejected or tabled cogeneration investments because they cannot find a way to control emissions adequately or to increase the cost savings to make the investment sufficiently attractive. Even if the emission standards could be met, the U.S. Environmental Protection Agency's nonattainment policy* creates a general uncertainty about future emission control requirements.

d. Government Regulation

Currently, the nature and extent of Federal Energy Regulatory Commission (FERC) and state public utility commission (PUC) jurisdiction over corporations that own or operate cogeneration facilities are not clearly defined. In some situations, FERC and PUCs clearly would not have jurisdiction under current law. In others, however, a clear regulatory policy does not exist.

Overall, industry management is therefore concerned about the potential regulation of cogeneration facilities that parallel, but do not transmit, power to the utility grid. Management is also concerned that industrial cogeneration plants may be required to deliver power to the grid to provide reserve or emergency capability for the utility and, thus, jeopardize manufacturing plant operations. In most instances, management responds to such uncertainty by increasing the required ROI.

III. COGENERATION DEVELOPMENT WITH FEDERAL GOVERNMENT ACTION

Federal actions to mitigate financial and regulatory constraints can stimulate between 20 and 40 trillion Btu of additional process steam cogeneration development by 1985 in the six industries examined. These actions correspond to less than 1% to more than 7% of the total process steam demand by 1985 (See Figure 4). The corresponding additional cogeneration electric capacity stimulated through government actions will range between 140 and 2600 MW and provide net annual energy savings of 5 to 94 trillion Btu.

A federal program consisting of some or all of these actions could provide up to about 200 trillion Btu of net annual energy savings by 1985. The corresponding additional cogeneration electric capacity stimulated by a federal program could total 6000 MW. This additional capacity is comparable to that expected to develop over the same period without federal action.

In the RPA study, the following actions were examined in terms of their potential effects on factors, including cogeneration development and resulting energy savings:

- A 30% investment tax credit
- A 12-year depreciation option
- Government loan guarantees
- Establishment of industrial electric power flat rate structures

---

* In a nonattainment area, a new source or a modification of an existing source that increases emission of a critical pollutant by more than 100 tons/year is subject to EPA's Emissions Offset Policy. The lowest achievable emission rate established by the operating facility of the same type must be met. All other company sources must be in compliance with emission standards. There can be no air quality degradation in the vicinity of the source, and there must be more than a one-for-one trade-off between new sources and the modified source, resulting in a new air quality benefit.

Evaluating the Effect of Specific
Government Actions

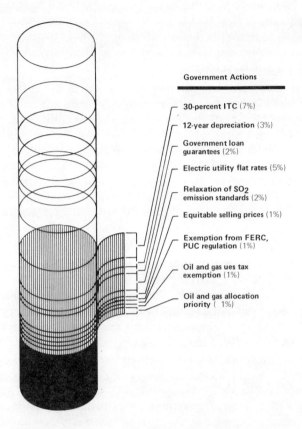

Government Actions

30-percent ITC (7%)

12-year depreciation (3%)

Government loan
guarantees (2%)

Electric utility flat rates (5%)

Relaxation of SO_2
emission standards (2%)

Equitable selling prices (1%)

Exemption from FERC,
PUC regulation (1%)

Oil and gas ues tax
exemption (1%)

Oil and gas allocation
priority (1%)

FIGURE 4. Evaluating the effect of specific government
actions.

- Relaxation of sulfur dioxide (SO_2) emissions standards
- Equitable prices for purchases of power by utilities from industrial cogeneration plants
- Exemption of industrial cogeneration plants operating *in parallel with, but not selling to, a utility* under Federal Energy Regulatory Commission (FERC) and public utility commission (PUC) jurisdiction
- Exemption from oil- and gas-use taxes
- Preferential treatment on oil and gas allocations

The financially directed actions would induce the greatest development of cogeneration and result in the largest energy savings. The remainder, most of them regulatory, will result in less energy savings.

A. Thirty-Percent Investment Tax Credit

The establishment of a 30% investment tax credit would increase cogeneration development by improving the financial returns. By 1985, the incremental cogeneration development resulting from the tax credit would exceed 400 trillion Btu of steam energy (or about 7% of the process steam demand in 1985) and 20 billion kWh of electric energy from an electric generating capacity increase of 2600 MW. In addition, cumulative energy savings of 720 trillion Btu are expected over the 1978 to 1985 period. This tax credit is purely a financial incentive. It has no regulatory or technical components. The action by Congress would require a federal tax code provision that would add 20% for new cogeneration equipment to the existing 10% new investment tax credit.

B. Twelve-Year Depreciation Option

Under federal tax codes, the depreciation now allowed is 22½ years on equipment. The establishment of a 12-year depreciation option for industrial cogeneration equipment, one of the most effective of the government actions considered, would increase cogeneration development by improving financial returns. This option would increase the level of cogeneration by an expected 160 trillion Btu (or about 2%) of steam cogeneration, 8.2 billion kWh of cogenerated electric energy, and 1050 MW of electric generating capacity. The expected cumulative energy savings of 295 trillion Btu is a function of the increase in electricity cogeneration. Like the 30% investment tax credit, the 12-year depreciation option is a purely financial incentive with no regulatory or technical components.

C. Government Loan Guarantee Program

A government loan guarantee program would increase cogeneration development by reducing the risk and, thus, the ROI hurdle rates. Through this action, the level of cogeneration would be raised by almost 100 trillion Btu of steam energy cogeneration (or about 2%), 4.7 billion kWh of electric energy cogeneration, and an electric generating capacity of about 620 MW. Cumulative energy savings resulting from an increase in electricity cogeneration are expected to be about 160 trillion Btu by 1985.

D. Establishment of Flat Electric Rates

After discussion with industry representatives, a benchmark estimate of 15% was established as the average increase in the price of electricity purchased from utilities by large industrial users if flat electric rates for these users were inaugurated. With this price increase, cogeneration would increase by improving the ROI and, thus, increasing the portion of technically suitable steam that would also be financially attractive for cogeneration. The resulting increase in cogeneration development would be 293 trillion Btu of steam cogeneration (or approximately 5%), 14.1 billion kWh of electric energy cogeneration, and over 1800 MW of electric generating capacity. By 1985, cumulative energy savings would be approximately 500 trillion Btu.

E. Relaxation of Sulfur Dioxide Emission Standards

Relaxation of SO_2 emission standards would improve the expected ROI for coal retrofit and new coal facilities by removing the requirement for installation of desulfurization equipment. Relaxation of standards from 1.2 to 3.0 lb/million Btu would allow operators of cogeneration plants to burn higher sulfur fuel without installing flue-gas desulfurization equipment or "scrubbers". However, operators of coal-fired industrial boilers would still be required to burn compliance coal or use scrubbers to meet air emission standards. In addition, regulations on other emissions would remain unchanged for both cogeneration plants and industrial boilers.

By 1985, this action would increase cogenerated steam by 126 trillion Btu (or about 2%), electric energy cogeneration by 5.9 billion kWh, and cogeneration capacity by 750 MW. The enhanced electricity cogeneration would result in cumulative energy savings of 170 trillion Btu by 1985.

F. Equitable Power Prices

Although there is no precise definition of the term equitable price, such a price would generally fall somewhere between the cogenerator's cost and the potential buyer's cost for power from an alternative source. Usually, when the sale price for the cogenerated power is about 10 to 20% below the purchase price for utility power, a cogeneration plant with capacity to sell power has approximately the same ROI as a plant that can only match industrial power needs.

The amount of cogeneration that would actually develop as a result of equitable sale prices, based on the economies of scale expected to result from the selling of power, is projected to be 410 MW of cogeneration capacity. Process steam cogeneration would increase by 54 trillion Btu in 1985 (or about 1%), and electric energy cogeneration would increase by 3.1 billion kWh.

Government action to ensure equitable prices for power purchased by utilities from industrial cogenerators would result in only moderate net energy savings of 13 trillion Btu in 1985 and cumulative savings of 100 trillion Btu. Moreover, the uncertainty about these savings is greater than the uncertainty about other actions considered because of the unresolved regulatory issues. Nonetheless, this action could be just as effective as the 30% investment tax credit and the 12-year depreciation proposals.

G. Exemption from FERC and PUC Jurisdiction

Management has expressed concern that an industrial plant might become subject to the authority of the state PUCs if there were a parallel interconnection with the grid, even if no cogenerated power were sold. Exemption of industrial cogeneration plants from government utility regulation would be the most effective response to this concern. Such action would require FERC and the state PUCs to relinquish all jurisdiction over industrial cogeneration plants, except perhaps during a declared state of emergency. However, it would, in no way restrict any federal or state authority over the sale of power from industrial cogeneration plants.

This policy option would increase cogeneration investment by reducing the ROI requirements, thus increasing the amount of process steam that is financially attractive. Although small, by 1985 the incremental increase in cogeneration development would be 35 trillion Btu of steam energy cogeneration (or approximately 1%), 1.6 billion kWh of electric energy cogeneration, and 220 MW of electric generating capacity. Cumulative energy savings would also be small, only 60 trillion Btu.

H. Exemption from Oil- and Gas-Use Tax

Congress is considering a tax on the use of oil and gas, imposed on industry beginning in 1979 and on utilities beginning in 1983. If cogeneration facilities were exempted from this tax, cogeneration development would increase by 73 trillion Btu of steam cogeneration (or approximately 1%), 3.6 billion kWh of electric energy cogeneration, and 460 MW of electric generating capacity. This measure would increase cogeneration investments by increasing cogeneration ROIs because of higher purchased electricity rates. This additional development would result in cumulative energy savings of 120 trillion Btu.

I. Oil and Natural Gas Allocation Priority

The compromise of a high oil or gas allocation priority in the event of an embargo would have relatively little effect on cogeneration development because industrial managers do not perceive the use of the standby petroleum allocation programs as a significant risk. The opportunity to receive a high allocation priority for cogeneration plants using oil would consequently affect the ROIs required for cogeneration investments only slightly. Moreover, although natural gas curtailments are of much greater concern to managers with respect to their current and near-term operations, very little natural gas would be used in cogeneration. For these reasons, increases in cogeneration development resulting from this government action, are only 22 trillion Btu of steam cogeneration (or less than 1%), 1 billion kWh of electric energy cogeneration, and 140 MW of an electric generating capacity in 1985. Cumulative savings would also be small, only 40 trillion Btu.

IV. COGENERATION DEVELOPMENT WITH ACTIVE INVOLVEMENT BY THE MAJOR PARTIES

Although federal actions can stimulate substantial additional development of industry owned and operated cogeneration plants, greater development can be achieved with direct participation by electric utilities. In the previous chapters, it was estimated that approximately 6000 MW of industry owned and operated cogeneration capacity would be installed by 1985 without federal action. With federal actions, this amount could reach almost 12,000 MW of additional cogeneration capacity. Recently, it has become apparent that many of the technical and investment constraints inhibiting industry investments in cogeneration plants could be mitigated if utilities participated in either the ownership or operation of such plants or both.

From greater industrial cogeneration development, the public sector could achieve significant benefits. With cogeneration, the demand for critical fuels such as natural gas and distillate oils for power production in industry and utilities would be reduced, thereby making these fuels more readily available to the remaining sectors, residential and commercial. Combined generation (cogeneration) of process thermal energy and electricity could provide significant energy savings and lower-cost electricity. Finally, environmental benefits could be realized through combined generation which can displace separate production of process thermal energy and electricity.

However, for the additional development of industrial cogeneration to approach the maximum levels practical, a cooperative effort must be undertaken by all four major parties involved, that is, the federal government, state governments, industry, and electric utilities. Because decisions made by each party influence cogeneration development, both active involvement by each party and coordination of efforts are needed to achieve optimum development. Of course, optimum development requires balancing the priorities and interests of each party involved. For example, the federal and state governments are concerned with government energy savings, coal conversion opportunities, electricity cost to users, quality of electric service, and environmental sensitivities. Industry, in contrast, is primarily concerned with dollar savings and regulation. Finally, interest of the electric utilities focuses on cost of electric power production, return on investment, capital requirements, and the capability to provide quality electric service.

A. Effects of Industrial Cogeneration on Electric Utilities

Utilities are concerned about industrial cogeneration, particularly about the loss of industrial customers who build cogeneration plants, and the interconnections of industrial cogeneration plants with the utility grid.

Significant industrial cogeneration development in a utility service area could severely affect the financial position of the utility. The loss of baseload customers could alter the utility electric load patterns so that the cost to the utility to produce electric energy for the remaining customers would increase. A greater portion of the remaining customers would be supplied by more expensive generating plants, i.e., from peaking and intermediate capacity plants. The revenue lost by the utility as a result of the loss of large customers, coupled with the increased cost per kWh of electric energy produced, could sharply reduce the financial return to the utility.

Interconnections between utility grids and industrial cogeneration plants also create concerns for utility managers. Generally, the utilities desire to have dispatching control over the electric power entering the grid to ensure system stability and security. If the amount of purchased power were significant, the utilities would want guarantees on time of delivery, amount, and length of the agreement, among other considerations.

In addition, virtually all utilities are concerned about government action that they believe recognizes only the industrial point of view, failing to take into account the total economics of the utilities and industrial companies and the effect on the various consumers of electric power.

B. Coordinated Actions

Direct utility participation offers the potential for greater industrial cogeneration development. A number of these industrial and utility concerns which limit development would be eliminated or greatly alleviated by utility ownership, joint ownership, or third-party ownership of cogeneration facilities. From industry's side, the fear of government regulation could be eliminated. Also, certain cogeneration investments could be economically attractive for a regulated utility that would not be attractive for an industrial company with higher ROI requirements. Further, where several industrial users are situated within a few miles of each other, greater economies of scale could accrue to multiple users from a larger plant. Finally, capital constraints would be greatly mitigated under joint or third-party ownership.

From the utility viewpoint, most of the concerns about industrial ownership would be alleviated if the utilities owned or participated in the operation of the cogeneration plants. However, there is one factor that would influence a utility's desire to produce excess power from a cogeneration plant to the grid. In some states, a utility-owned plant that supplied power only to on-site customers is exempt from regulation, thus allowing the utility to receive a greater-than-regulated rate of return. Therefore, depending on the cost of excess power relative to other sources of power, the utility may have no incentive to build a plant that supplies excess power.

Additional development with a greater utility role can be achieved through cooperative actions taken by the major interacting parties. The following are some of the more important actions that each of the parties should take as part of a cooperative effort to produce a combination of conditions facilitating cogeneration plant investment. Federal actions were discussed in the previous chapter, see Figure 5.

1. State Government Actions

Over time, state government actions should prove to be as important as, if not more important than, federal action. This is because the regulation of electricity rates is under the direct control of state agencies. One important action that can be taken by states is to review rate structures and to consider changes that could favor cogeneration development.

Other possible state actions are similar to various possible federal actions, addressed previously, in such areas as balancing energy and environmental policies at the state level, clarifying PUC jurisdiction, and providing financial incentives. A significant step for consideration is the exemption of all cogenerators, both industry owned and utility owned, from regulation. This would have a large impact on cogeneration development and could provide lower-cost electricity to the consumer.

**Opportunities for Greater
Cogeneration Development**

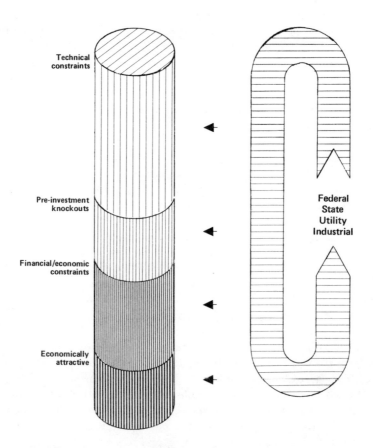

Technical
constraints

Pre-investment
knockouts

Financial/economic
constraints

Economically
attractive

Federal
State
Utility
Industrial

FIGURE 5.   Opportunities for greater cogeneration development.

2. Industry Actions

Industry can consider four actions that could change their decision on cogeneration from negative to positive.

First, there should be a willingness to negotiate with electric utilities. Many constraints to cogeneration, including fear of possible government regulation of the cogeneration facility, can be eliminated through partial or full utility ownership.

Second, industry should accept, and work within the context of, a substantial government presence in the energy field. Whatever one's philosophical view on free enterprise may be, it is clear that energy is too precious and too important to our economy and our society for there to be any expectation that the government will ever again assume a relatively hands-off posture. In fact, where energy matters involve many industries as well as the public as a whole, there is no alternative to government mechanisms for the ordering of priorities.

Third, many industrial companies should explore other technical plant configurations and other ownership options in analyzing cogeneration plant investment alternatives.

Fourth, many companies underestimate the likely future increase in electricity prices and fuel prices in calculating the ROI of a cogeneration plant. While higher energy prices do not always favor a decision for cogeneration investment, exploring a range of potential electricity and cogeneration fuel prices will make for a better investment decision.

3. Electric Utility Actions

Mirroring these suggestions to industry, one can suggest to utilities that they enter willingly into negotiations with industrial companies. One may also suggest that they accept a government role in energy price and supply.

It is further believed that utilities could make a fundamental mistake if they do not compare the cost of cogenerated power with the cost per kWh of new electric power plants. Further, utilities should study carefully the economics of small plants, which in total can constitute a significant portion of some electric utilities' service area capacity.

Finally, utilities should calculate standby power charges, or other rates for backup capacity, on the basis of the total number of plants served, and not charge for full standby capacity for each individual manufacturing plant served.

REFERENCES

1. Resource Planning Associates, The Potential for Cogeneration Development in Six Major Industries by 1985, HCP/M60172-01/2, Resource Planning Associates, Cambridge, Mass., for the Federal Energy Administration, Washington, D.C., December 1977.
2. Spencer, R. S., Energy Industrial Center Study, PB-2431 823/2ST, Dow Chemical Company, Midland, Mich., for the National Science Foundation, Washington, D.C., June 1975.
3. Nydick, S. E., et al., A Study of In-plant Electric Power Generation in the Chemical, Petroleum Refining and Paper and Pulp Industries, PB-2551 659/5ST, Thermo Electron Corporation, Waltham, Mass., for the Federal Energy Administration, Washington, D.C., July 1976.
4. Hatsopoulos, G.N., Widmer, T. F., Gyftopoulos, E. P., and Sant, R., A National Policy for Industrial Energy Conservation, Thermo Electron Corporation, Waltham, Mass., April 1977.
5. Barnes, R. W. and Whiting, M., Demand Conservation Panel Report to the Committee on Nuclear and Alternate Energy System, National Research Council, National Academy of Sciences, Washington, D.C., 1976—77.
6. Ross, M. H. and Williams, R. H., The potential for fuel conservation, Technol. Rev., 79(4), 48, 1977.
7. Karkheck, J., Beardsworth, E., and Powell, J., Technical and Economic Aspects of U.S. District Heating Systems, BNL-21287, Brookhaven National Laboratory, Upton, N. Y., for the Energy Research Development Administration, Washington, D.C., April 1976.
8. Karkheck, J., Powell, J., and Beardsworth, E., Prospects for district heating in the United States, Science, 195, 11, 1977.
9. Mixon, W. R., et al., Technology Assessment of Modular Integrated Utility Systems, ORNL/HUD/MIU5-24, Oak Ridge National Laboratory, Oak Ridge, Tenn., for the Department of Housing and Urban Development, Washington, D.C., 1976.
10. Barnes, R. W., A Comparative Evaluation of Recent Reports on the Energy Conservation Potential from Cogeneration, ORNL/TM/66-02, Oak Ridge National Laboratory, Oak Ridge, Tenn., for the U.S. Department of Energy, Washington, D.C., 1979.

Chapter 16

POTENTIAL FUTURE MARKET FOR DISTRICT HEATING AND
TOTAL ENERGY SYSTEMS

M. Olszewski, M. A. Karnitz, and W. R. Mixon

TABLE OF CONTENTS

I. INTRODUCTION

Space and water heating in the residential-commercial sector accounts for about 20% of the national energy budget. This represents an annual fuel consumption of about 15 quads*/year. Residential-commercial space cooling (air conditioning) accounts for an additional 4% of the national energy budget. Thus, space conditioning and water heating in this sector accounts for about one fourth of U.S. energy consumption. Furthermore, over 90% of these heating needs are satisfied using oil and natural gas, while electricity is the major energy source for air conditioning.

Given the current U.S. energy situation, it would be desirable to substitute plentiful domestic supplies of coal and uranium for the scarce natural gas and oil fuels currently in used for space heating applications. An alternative option, which would also be highly desirable, would be to utilize these scarce fuels more efficiently to satisfy these heating requirements.

Large-scale, utility-owned district heating grids appear to be a promising alternative for accomplishing the former objective, while smaller-scale, "stand-alone" Total Energy systems are an attractive technology for accomplishing the latter.

District heating has received increased attention during the past few years for a number of reasons. In addition to the flexibility in fuel requirements mentioned above, dual-purpose stations can be used in the system to increase the thermal efficiency of generating electric power. Thus, district heating provides a means of utilizing abundant coal and nuclear resources and allows them to be utilized in an efficient manner.

Total Energy systems have been used for commercial and institutional applications for a number of years. Because of their efficient fuel-use characteristics, they have also received attention for multifamily residential applications. These systems are much smaller in scale than the district heating systems discussed above and generally supply the electric and heat needs for industrial and commercial applications.

Because there are a number of inherent differences between the two systems, the market potential for district heating and Total Energy systems will be explored separately.

II. DISTRICT HEATING

District heating has been used in many parts of the world and has been widely accepted in Northern Europe. These systems typically supply steam or hot water to residential and commercial customers, although industrial process applications are sometimes included.

District heating was introduced in the U.S. in the late 19th century. Growth of these systems was based on steam delivery to commercial customers located in downtown areas. In contrast, European system development occurred after World War II. These systems used hot water for heat transport and served commercial and residential customers. Because of the high losses associated with steam energy piped over long distances, hot water has proven to be more economical for large systems. Lower transmission costs have allowed for more expansive growth along with the use of remotely located dual-purpose plants. District heating, of course, does not necessarily have to include dual-purpose plants. Many of the existing U.S. systems, in fact, utilize heat-only plants to supply thermal energy. However, the conservation and economic advantages of district heating are greatly improved when dual-purpose plants are used.

---

*   1 quad = 10^{15} Btu.

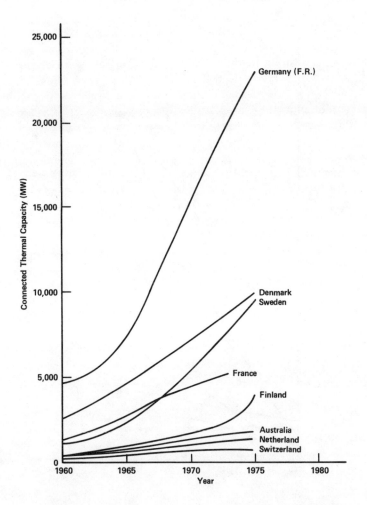

FIGURE 1.   Development of connected thermal capacity (Western Europe).

Because of the considerations mentioned above, our market projections will concentrate on systems that utilize hot water transport. The analysis will further be confined to systems that include dual-purpose plants to supply a portion of the system thermal demand.

A. Background

Before examining the market potential for district heating in the U.S., it will be beneficial to briefly review the status of district heating in the U.S. and Europe. Since district heating has been successful in Europe, a comparison of U.S. market projections and current European implementation levels will allow these projections to be viewed from a realistic perspective.

As shown in Figure 1, the growth of district heating in Western Europe has been substantial for a number of countries during the past 15 years.[1] At the present time, district heating satisfies an average of 7% of the residential heating needs of the countries included in Figure 1.

The growth of district heating in Eastern Europe[1] has been even more dramatic. As shown in Figure 2, the Soviet Union is the world's leader in district heating implementation, with over 50% of the domestic heat demand being satisfied by district heating.

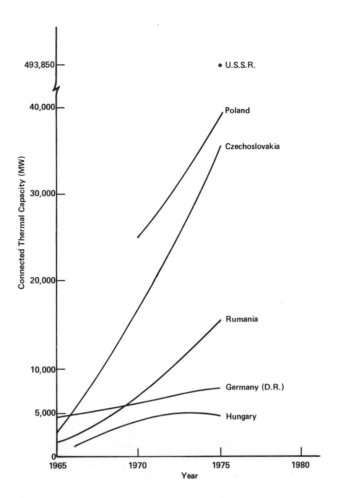

FIGURE 2.   Development of connected thermal capacity (Eastern Europe).

The steady growth exhibited in Figures 1 and 2 is expected to continue, with large scale expansion expected in the near future. West Germany is now considering a plan for a national heating grid. This grid would supply all towns with a population of 40,000 or more. Estimates indicate that by the year 2000 about 60% of the total heat load in Finland will be supplied by district heating. Expansion plans for other European countries call for 25 to 50% of the domestic heat load to be satisfied by district heating systems within the near future.

The thermal and economic benefits of dual-purpose stations were recognized during the early stages of district heating growth in Europe. Thus, many of these systems make extensive use of dual-purpose stations. For example, about two thirds of the total district heating supply in West Germany is provided by over 100 dual-purpose stations. In the Soviet Union, over 1000 dual-purpose stations are in operation in about 800 cities. These stations satisfy over 30% of the USSR's power generation needs and over 50% of the domestic heat demand. Indications are that dual-purpose plants will play an even more important role in the expansion of European district heating systems.

In contrast to the European annual growth rate of about 20%, U.S. district heating has been expanding at a rate of about 5% per year over the last 10 years. In 1975, total utility steam sale for heating[2] was about 40×10^9 kg. Estimates[3] of nonutility (government institutions, college campuses, etc.) steam producton indicate a total quantity about equal to the utility sales figure. Therefore, the total amount of steam used for district heating in 1975 was on the order of 82×10^9 kg. District heating thus satisfied approximately 1% of the demand for heating in the U.S. Utility statistics[2] also indicate that 313×10^6 kg of steam were sold to provide about 72,610 kW of refrigeration for chilled-water production.

As previously discussed, U.S. systems do not make extensive use of dual-purpose systems. Of the total utility steam delivered in 1975, only 35% was supplied from back-pressure or extraction turbines. Thus, less than 0.5% of the U.S. residential-commercial demand for heating in 1975 was satisfied by dual-purpose stations.

It is clear from the previous discussion that significant expansion of district heating can occur in the U.S. before European implementation levels are achieved.

B. Estimates of National Implementation Levels

Recent analysis of the market potential for district heating in the U.S. has focused on two analysis techniques. The Brookhaven study[4] focused on a comparative U.S.-European population density method, and concluded that 50 to 55% of the U.S. population could be served by district heating. A second study[5] correlated existing heat sources with population centers and estimated that 15% of U.S. space heating needs could be economically converted to be served by district heating systems. Since there is considerable variation between the two estimates, a review of the details of both approaches will be helpful in determining which estimate represents a realistic upper bound on U.S. implementation.

The Brookhaven work compared the annual per capita residential space heating consumption of various American cities with European cities having extensive district heating systems. Based on their data, shown in Table 1, they concluded that U.S. and European demands are roughly equivalent. It is further assumed that the hourly wage rate comparison indirectly reflects district heating equipment and installation costs. Based on equivalent European energy demands and material costs, the study concluded that many American cities were suited for district heating.

Proceeding on this conclusion, they attempted to define regions in the U.S. where economic conditions were favorable for district heating by modeling nine representative U.S. regions. The model incorporated population density, climate, total population, labor and material costs, and insulation and floor space. Their results indicated favorable economic conditions for all regions.

The next step in the analysis was to estimate national implementation levels. For this task, average values for degree-day data and installation costs were used. Figure 3 shows the total population levels that are serviceable as a function of heat cost. The study concluded that at a heat cost of $4.00/GJ (the assumed average heat cost) about 50% of the population (100 million people) could be served by district heating.

While the Brookhaven approach is an excellent method, there is some concern that cost estimates used in the study were too low. The cost for heat from cogeneration plants, for example, was based solely on the additional fuel costs. Capital costs associated with plant modifications for retrofitting existing power stations were neglected. Additional capital costs, above those of a standard condensing power station, were similarly neglected for new stations. Based on experience from the Minneapolis-St. Paul study,[6,7] described in the next section, it also appears that the distribution network costs are low.

Adjustment of the cost data could have a significant impact on the district heating implementation curve in Figure 3. Based on the results of other studies,[5-7] the higher

TABLE 1

A Comparison Among Several European Cities Known to have Extensive District Heating Systems and Similar American Cities

City	Climate[a]	Total population (thousands)	Population density (people/km²)	Annual per capita residential space heat consumption (GJ)	Hourly wage (U.S. $)[b]
Stockholm	8100	750	4015	23⎤	7.12
Malmo	6700	254	2340	19⎦	
Helsinki[c]	8400	750	600	18	NA[d]
Hamburg	6300	1800	2390	21⎤	6.19
West Berlin	6100	2000	4170	20⎦	
Denver	6300	515	2080	34	
Chicago	6200	3367	5840	33⎤	
Minneapolis	8400	434	3040	45⎱	6.22
Detroit	6200	1511	4230	33⎰	
Buffalo	7100	463	4330	38⎦	
New York	5000	7895	10200	27	

[a] Calculated as degree-days, cumulative average daily variation below 65°F (18°C).
[b] 1975 National hourly types, including benefits (from Reference 13).
[c] Helsinki metropolitan area.
[d] NA, not applicable.

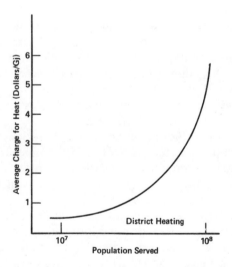

FIGURE 3.   Comparison of net (after conversion) heat costs (fuel only, excludes capital for end-use devices).

cost levels would tend to make district heating options less competitive with alternate heating sources. Thus, the potential for implementation would be decreased and the implementation curve in Figure 3 would be shifted to the left. Thus, the rapid increase that occurs at a population of 100 million would be shifted to a lower population.

A more recent estimate of the potential of district heating by Argonne National Laboratory[5] uses a somewhat different approach. Their investigation correlated existing heat sources with population centers. These results showed that there are 337 exist-

TABLE 2

Application of District Heating Systems (DHS) to the Residential/Commercial Sectors —
Study Results for Most Attractive Cities

Urbanized area	Total potential scarce fuel thermal demand savings from DHS (10^{12} Btu)	Cumulative total	Total potential scarce fuel thermal demand savings from DHS (%)	Cumulative %
1. New York City — New Jersey[a]	287	287	14	14
2. Washington, D.C.	216	503	11	25
3. Chicago	178	681	9	34
4. Boston[a]	167	848	8	42
5. Detroit[a]	130	978	6	48
6. San Francisco[a]	114	1092	6	54
7. Philadelphia[a]	96	1188	5	59
8. Baltimore[a]	84	1272	4	63
9. Providence, R.I.	56	1328	3	66
10. Milwaukee[a]	42	1370	2	68
11. Cleveland[a]	37	1407	2	70
12. Buffalo	36	1443	2	72
13. Minneapolis[a]	35	1478	2	74
14. Springfield, Mass.	32	1570	2	76
15. Louisville, Ky.	28	1538	1	77
16. Wilmington	26	1564	1	78
17. Pittsburgh[a]	23	1587	1	79
18. Indianapolis[a]	21	1608	1	80
19. St. Louis[a]	21	1629	1	81

[a] Cities with existing steam systems.

ing power plants within 10 mi of urban areas having more than 50,000 residents. Therefore, they limited their implementation analysis to urban areas having existing power
plants within a 10 mi radius. They assumed only the intermediate- and base-load plants
would be modified, thus eliminating all peaking plants from further consideration.
The potential for new plants, however, was included.

The district heating market potential was then estimated by performing a marginal
cost analysis. The cost of delivered electricity was assumed to be fixed at 2.5¢/kWhe.
If thermal energy could be delivered at a cost less than or equal to the assumed electric
cost, then the area under consideration was judged to be a potential district heating
service area. Thermal energy at the plant gate was valued in terms of lost electrical
generation. For one unit of electricity sacrificed, there was a gain of 3.75 units of
thermal energy. Therefore, the cost of thermal energy at the plant gate was 0.67¢/
kWht. This yielded an allowable thermal transport cost of 1.8¢/kWt (i.e., 2.5¢ minus
0.67¢). Using this transport cost, an allowable service area was defined, and the scarce
(oil and gas) fuel savings determined.

The results given in Tables 2 and 3 show the scarce fuel savings for urban areas and
regions, respectively. The total amount of scarce fuel that could be replaced by district
heating in urban areas was 2×10^{15} Btu/year, or approximately 15% of the U.S. demand for space and water heating.

TABLE 3

Application of District Heating Systems (DHS) to the Residential/Commercial Sectors — Results by Region and U.S. Total

Region	Urbanized area[a] total annual thermal demand supplied by S.F. (10^{12} Btu)	Total urbanized area S.F.[b] thermal demand that could be supplied by DHS (10^{12} Btu)	Scarce fuel savings potential by retrofit of N.S.F.[c] plants (10^{12} Btu)	Scarce fuel savings potential by retrofit of S.F. plants (10^{12} Btu)	Total scarce fuel savings potential by retrofit (10^{12} Btu)	S.F. thermal demand that could be supplied by new plants (10^{12} Btu)
New England	535	347	18	24	42	281
Mid-Atlantic	1201	501	134	60	194	272
E. North Central	1225	529	401	24	425	211
W. North Central	369	87	72	4	76	27
S. Atlantic	719	364	50	23	73	278
E. South Central	156	39	51	0	51	1
W. South Central	389	8	0	4	4	0
Mountain	145	21	16	3	19	4
Pacific	572	114	0	10	10	93
U.S. Total	5311	1009[d]	742	152	894	1167

a Urbanized area defined as center city >50,000 and population density >1000/mi². Only urbanized areas with power generating plants within 10 mi. of border considered.

b S.F., scarce fuels (oil and natural gas).

c N.S.F., nonscarce fuels (coal and nuclear).

d Planned nonscarce fuel power generation capacity within 20 mi of urbanized areas between 1974 and 1984 is 227,000 MWt. At a 0.6 load factor, 4072 × 10^{12} Btu/yr could be supplied to these areas which is sufficient to more than meet potential demand.

A comparison of the two studies indicates that the Argonne work yields the most realistic estimate for the potential for district heating in the U.S. Therefore, 15% of the space and water heating demand appears to be a realistic upper bound. If implemented at this level, the total U.S. district heating thermal capacity would be approximately 300,000 MW. Comparing this number to the current European implementation levels illustrated in Figures 2 and 3 indicates that this level is a realistic estimate, especially in view of the ambitious European expansion plans currently under consideration.

C. District Heating Potential for the Minneapolis-St. Paul Area

A large-city district heating study for the Minneapolis-St. Paul metropolitan area[6] provides some detail concerning how such levels of implementation could be achieved for a particular region. A preliminary report[7] indicates favorable economics for a large system, which would include commercial and residential areas, if advanced temperature-resistant plastic-pipe technology is used. For conventional piping technology, the economics appear favorable only for the commercial central core area.

Estimates indicate that the service area would include about 900,000 people and have a mean heat load density of 11 MWt/km². The service area would consist of the city core area, which covers about 26 km² and has a heat load density of 56 MW/km², and surrounding regions with population densities greater than 1600/km². These surrounding regions have an average heat load density of 7.6 MWt/km². The peak heat load for the system would be about 4200 MWt.

The joint U.S.-Swedish study[7] outlines a three-stage approach for developing the system, which would take about 15 years. Initially, an area of only 2 to 3 km² of the main business district would be served. Portable boilers would be used as the energy source during this stage of development. The next stage would expand the system to the entire 26 km² Twin Cities core area. The heat load for this area is about 1500 MWt. During this stage, two units at two different power plants would be modified for dual-purpose use. These units would be used as base-load units and provide about 600 MWt. The final stage would be the connection of the residential areas. This would raise the system heat load to 4200 MWt.

Approximately half (2200 MWt) of the peak heat load would be supplied by the base-load dual-purpose units, and the remaining 2000 MWt would be supplied by inexpensive heat-only units. The dual-purpose base-load units would supply 90% of the total annual energy load, while heat-only units would supply about 10%.

The estimated fuel use for the district heat system is 2.21×10^7 GJ/year. This compares to 6.05×10^7 GJ/year for the same area supplied by conventional boilers and furnaces. Thus, implementation of district heating would result in a 63% reduction in fuel consumption. The difference in scarce fuel consumption between the two options amounts to a savings of 5.22×10^7 GJ/year.

Results of the Twin Cities study indicate that significant fuel savings are possible, especially if advanced piping technology is used. They also indicate that the likelihood of connecting residential areas to the district heating system depends on the availability of low-cost piping technology and the future price of natural gas.

III. TOTAL ENERGY SYSTEMS

Historically, Total Energy (TE) has been used in residential, commercial, and institutional market sectors as systems which normally generate all electricity required by the users. System operation, and the availability of waste heat, follows electric demand, although emergency power connections with the local electric utility may be used.

Analytical studies of Modular Integrated Utility Systems (MIUS), which includes TE, also stressed applications that were independent of conventional utility systems.[8-11] Reliability of service and efficient operation at part load were provided by the use of multiple components and excess capacity. It was recognized that TE, MIUS, and other community systems concepts could realize significant advantages in system cost, efficiency, and maintenance if implemented as an integral part of the electric utility, but assessments were based on "stand-alone" systems because of:

1.   The history of TE applications
2.   Institutional factors and utility company attitudes in the early 1970s
3.   The use of a conservative approach by assessing MIUS applications in what was probably a worst case

The estimate of the potential TE market discussed in this chapter was based on the use of systems that were independent of the electric utility grid in multifamily residential-commercial developments. Identification of potential market was constrained by several technical and economic characteristics of the systems. The consideration of a stand-alone system implies that the TE plant would probably not be owned by an electric utility company and would, therefore, be planned and installed by others as an integral part of the community to be served. Considering the need for central planning, the most appropriate market would be housing developments or other communities under single ownership or management. The significant cost of distributing thermal energy from a central TE plant to each building served, and the added cost penalty of installing underground piping systems in established neighborhoods, further limits the market to new communities of medium to high load density. The economic viability of TE depends on many site-specific conditions, but studies indicate that stand-alone systems that include thermal distribution to many buildings would have to be large enough to serve the equivalent of several hundred dwelling units. Thus, one appropriate market sector consists of new multifamily housing and associated commercial buildings that are part of a large development project.

A. Potential Applications

One broad category of TE and MIUS application is the long-range, planned use of the systems for providing some fraction of the country's new utility capacity over the next two or three decades. Potentially attractive applications would include the following:

Urban redevelopment — In high-density urban redevelopment areas, the existing utility infrastructure would likely be outdated and inadequate to meet increased service demands, and new conventional utility construction could be prohibitively expensive or disruptive. For example, treated liquid waste from MIUS could be discharged to storm drains, thereby relieving overloaded conventional sewers and treatment plants.

New town and outlying developments — TE would be applicable to serve the high-density centers of new towns, suburban or fringe-type developments (such as in new communities, large multifamily developments, or planned unit developments), shopping and other commercial centers, industrial developments, etc. District piping systems would be economical in high-density areas, and the fact that MIUS can be completely independent of conventional utilities saves the expensive extension of conventional utility transmission lines to outlying areas. Surrounding medium- and low-density areas could also be served by the MIUS installation, but heating and cooling would probably be provided by individual building systems, such as electric heat pumps.

Institutions — TE applications may include university campuses, medical centers, and hospitals. The Department of Health, Education, and Welfare, in response to national policy for use of energy-conserving concepts, assisted in the transfer and evaluation of MIUS technology by conducting conceptual design and feasibility studies for integrated utility installations at three universities.[12-14]

Contingency situations — Another category of MIUS application is the provision of service in contingency situations. Examples would be use in communities where local restrictions on waste treatment or other utilities prevent construction of housing, and in isolated communities, both new and existing, which are located too far from municipal-sized utility plants for economical transmission and services. Such applications could include the following:

- Second home, resort, and vacation developments serving relatively small, affluent populations are suitable. These areas are generally environmentally sensitive and have little or no existing infrastructure.
- Military bases and other government installations are suitable. In keeping with a national policy for energy conservation, the Department of Defense is examining the feasibility of using TE systems for every project above a particular size.
- Remote villages, such as Alaskan villages and Indian reservations, where no utilities exist, are suitable.

B. Potential Residential Market

Historical patterns of residential construction and projections of future housing demands were examined to determine the potential TE market in residential applications.[11] Data on housing construction starts for the years 1965 to 1973 indicate an increasing trend in construction of multifamily units. Of the total of private- and public-owned starts, different sources showed multifamily units to be 44 to 46% in 1970 and 45 to 55% in 1973.

Future housing demands are difficult to estimate because of uncertainties in birth rates, migration patterns, and the rate of replacement of units. In addition, there is no certainty that new housing will actually be constructed in accordance with projected needs. Published projections indicate a total housing demand of almost 3 million units/year, and an average demand for multifamily housing of about 38% of the total (excluding mobile homes) in the years 1975 to 2000. The projected decreasing demand for multifamily housing from 1975 through 1990 was not consistent with the 1970 to 1973 actual construction trends. Both sets of data were considered, but rapidly increasing costs for land and single-family housing, indications of more favorable opinions toward multiunit housing, and current construction data all suggest a trend to increasing multiunit construction.

The projected number of medium- to high-density residential-commercial projects and the distribution by project size are essential data which could not be found in the literature. As a part of MIUS program activities, surveys were completed on existing and new multifamily developments in Houston (by NASA) and on HUD-insured mortgages in 1970 in order to determine the size distribution of multifamily projects. The results, shown in Table 4, indicate considerable divergence for projects containing 150 or more dwelling units. HUD insurance data indicate a considerable percentage of the largest projects, but HUD-insured housing represents only a part of new construction. Houston has a high percentage of large developments and new multifamily construction and is even more heavily weighted toward large projects, but Houston may not represent a typical metropolitan area.

Based on trends of current construction starts, it was assumed that 50% of total

TABLE 4

Multifamily Developments in 1970

Multifamily project size (number of units/ project)	Projects (%)		
	HUD insured (1970)	Houston construction (1970)	Existing Houston (1970)
50 or more	93	94	88
100 or more	72	85	74
150 or more	52	72	59
200 or more	28	60	45

housing would be multifamily. Total housing could range from 2,000,000 units/year, based on current construction, to a projected demand of almost 3,000,000 units/year. Of multifamily construction, a range of about 28 to 50% could be in single projects large enough for MIUS application (Table 4).

Thus, the MIUS market could consist of 280,000 to 500,000 units/year based on total construction rates, and 420,000 to 750,000 units/year based on projected total demand. The potential MIUS market was estimated to be 400,000 units/year, which is less than the average of the two extremes and is consistent with the combinations of (1) the high annual demand and the low number (28%) of large projects and (2) the current low construction rate and the high number (50%) of large projects.

There are many factors that could make the market higher or lower. Using 50% of the total housing as multifamily may be high, but many multiunit projects include single-family units which could also be served by MIUS. Projects containing only 200 units may be too small for the economical use of MIUS, but many projects would have associated commercial or public buildings which significantly increase total service demands. Considering these associated buildings, the market could almost triple if projects having 100 housing units are feasible, or could be much smaller if projects significantly above 200 units are required.

C. Potential Impact

The amount of installed generating capacity required to serve the potential market of 400,000 dwelling units (DU) per year varies with the type of utility and HVAC systems in use. If represented by a garden-apartment model and Philadelphia weather, 400,000 DU would have peak electric demands of about 900 MW when using TE, but this would displace demands on the conventional electric utility of about 2,500 MW if heat pumps or electric resistance heating is used, and about 1,300 MW if buildings use central boilers and electric air conditioners. Thus, as an upper limit, complete penetration of the appropriate residential market might displace the need for 2,000 MW of central station generating capacity per year.

Comparative systems studies described in Chapter 6 indicate that the use of TE would conserve primary fuel energy, but the oil- (or gas-) fueled TE systems would consume more oil (or gas) than would conventional systems.

Based on an overall average of results for five cities and five conventional HVAC models, the use of TE was estimated to save 30% of primary fuel energy. For the maximum residential market of 400,000 DU/year, energy savings would amount to about 19×10^{12} Btu/year.

D. Other Potential Markets

The *1972 Total Energy Directory and Data Book*[15] provides data on 508 installations in the U.S. Of these installations, 46% were industrial, 34% commercial, 14% institutional, and 6% apartment complexes. No current TE surveys have been completed, but increasing interest in cogeneration and the organization of the International Society for Cogeneration[16] have prompted the planning for the establishment of a cogeneration data base. The 1972 distribution of installations, however, indicates that there could be a substantially greater potential market in commercial and institutional applications than in apartment developments, perhaps eight times the residential market.

E. Market Factors

The potential market and actual use of TE is strongly influenced by a variety of technical, institutional, and economic factors. The preceding market estimate was based on the use of stand-alone systems with the additional capital cost of multiple components and excess capacity required to ensure reliability. Full integration of TE plants with the electric utility grid would reduce costs, make TE competitive in smaller communities, and increase the potential market. Utility interties should be strongly encouraged by the establishment of reasonable electric power interchange rates that include credits to the cogeneration plant for the firm availability of capacity.

Several ways of promoting the use of cogeneration have been suggested[17-18] that would improve the economics and increase the market for TE. These include:

- Additional tax credits, which may be of more importance to smaller installations
- Exemption from federal and state public utility regulation
- An efficiency-weighted oil and gas use tax
- Promotion of utility ownership
- Resolution of conflicts with environmental regulations
- Exemption from coal conversion
- Elimination of declining block electric rates
- Favorable treatment with respect to fuel allocations and curtailment

Nonindustrial cogeneration concepts offer a significant potential for energy conservation and considerable flexibility in the way to best provide services to any specific community. They are not proposed as a complete substitute for conventional utilities, but there are a significant number of applications for which community systems would be of public benefit and should be considered as an option.

REFERENCES

1. **Beresovski, T. and Spiewak, I.,** Urban district heating using nuclear heat — a survey, *At. Energy Rev.,* to be publishd.
2. **Anon.,** Statistical Committee Report of Industry Statistics for 1975, paper presented at the 1976 Int. District Heating Assoc. Meet., Saratoga Springs, New York, June, 1976.
3. **Clymer, E.,** personal communication with M. Olszewski, Oak Ridge National Laboratory, Oak Ridge, Tenn., January 1976.
4. **Karkheck, J., Power, J., and Beardsworth, E.,** The Technical and Economic Feasibility of U.S. District Heating Systems Using Waste Heat from Fusion Reactors, BNL-50516, Brookhaven National Laboratory, Upton, New York, February 1977.
5. **Anon.,** Potential for Scarce Fuel Saving in the Residential/Commercial Sector Through the Application of District Heating Schemes, Argonne National Laboratory, Argonne, Ill., 1977.
6. **Karnitz, M. A. and Rubin, A.,** Large City District Heating Studies for the Minneapolis-St. Paul Area, ORNL/TM-6283, Oak Ridge National Laboratory, Oak Ridge, Tenn., May 1978.

7. **Margen, P., Cronholm, L., Larsson, K., and Marklund, J.,** Overall Feasibility and Economic Viability for a District Heating/New Cogeneration System in Minneapolis-St. Paul, ORNL/TM-6830/P3, Studsvik Energiteknik AB, for Oak Ridge National Laboratory, Oak Ridge, Tenn., October 1979.

8. **Samuels, G., Robertson, R. C., Boegly, W. J., Jr., Breitstein, L., Grant, R. E., Griffith, W. C., Meador, J. T., Miller, A. J., Payne, H. R., and Segaser, C. L.,** MIUS Systems Analysis — Initial Comparisons of Modular Integrated Utility Systems and Conventional Systems, ORNL/HUD/MIUS-6, Oak Ridge National Laboratory, Oak Ridge, Tenn., June 1976.

9. **Hise, E. C., Boegly, W. J., Jr., Kolb, J. O., Meador, J. T., Mixon, W. R., Samuels, G., Segaser, C. L., and Wilson, J. V.,** MIUS Systems Analysis — Comparison of MIUS and Conventional Utility Systems for an Existing Development, ORNL/HUD/MIUS-20, Oak Ridge National Laboratory, Oak Ridge, Tenn., June 1976.

10. **Fulbright, B. E.,** MIUS Community Conceptual Design Study, NASA TM X-58174, National Aeronautics and Space Administration, Lyndon B. Johnson Space Center, Houston, Texas, June 1976.

11. **Mixon, W. R., Ahmed, S. B., Boegly, W. J., Brown, W. H., Christian, J. E., Compere, A. L., Gant, R. E., Griffith, W. L., Haynes, V. O., Kolb, J. O., Meador, J. T., Miller, A. J., Phillips, K. E., Samuels, G., Segaser, C. L., Sundstrom, E. D., and Wilson, J. V.,** Technology Assessment of Modular Integrated Utility Systems, Vol. 1, Summary Report, ORNL/HUD/MIUS-24, Oak Ridge National Laboratory, Oak Ridge, Tenn., December 1976.

12. **Kirmse, D. W. and Bronn, Carl, Jr.,** Integrated Utility Systems Feasibility Study and Conceptual Design at Central Michigan University, HEW Contract 100-75-0181, U.S. Department of Health, Education, and Welfare, Washington, D.C., October 1976.

13. **Kirmse, D. W. and Manyimo, S. B.,** Integrated Utility Systems Feasibility Study and Conceptual Design at the University of Florida, HEW Contract 100-75-0181, U.S. Department of Health, Education, and Welfare, Washington, D.C., December 1976.

14. **Paul L. Geringer and Associates,** Community Application of Integrated Energy/Utility Systems — Report No. 4, Conceptual Design, HEW Contract 100-77-0014, U.S. Department of Health, Education and Welfare, Washington, D.C., October 1977.

15. *Total Energy, 1972 Directory and Data Book,* Vol. 9 (No. 1), Total/Energy Publ. Co., San Antonio, Tex., January, 1972.

16. **Anon.,** Press Release of May 5, 1978, Personal communication from Int. Cogeneration Society, Inc., 800 17th Street, N.W., Suite 701, Washington, D.C.

17. **Solt, J. C.,** Fuel Savings Through the Use of Cogeneration, Solar Division of Int. Harvester, personal communication.

18. **Jovetski, John,** Cogeneration: Washington fiddles while industry burns, *Power,* 122, 35, 1978.

Index

INDEX

D

E